Johannes Schuler

The Economics of Soil Conservation Policies in Germany

Johannes Schuler

The Economics of Soil Conservation Policies in Germany

An Economic Analysis of the Implementation Options of Soil Conservation Policies in Brandenburg/Germany

Südwestdeutscher Verlag für Hochschulschriften

Impressum/Imprint (nur für Deutschland/ only for Germany)
Bibliografische Information der Deutschen Nationalbibliothek: Die Deutsche Nationalbibliothek verzeichnet diese Publikation in der Deutschen Nationalbibliografie; detaillierte bibliografische Daten sind im Internet über http://dnb.d-nb.de abrufbar.
Alle in diesem Buch genannten Marken und Produktnamen unterliegen warenzeichen-, marken- oder patentrechtlichem Schutz bzw. sind Warenzeichen oder eingetragene Warenzeichen der jeweiligen Inhaber. Die Wiedergabe von Marken, Produktnamen, Gebrauchsnamen, Handelsnamen, Warenbezeichnungen u.s.w. in diesem Werk berechtigt auch ohne besondere Kennzeichnung nicht zu der Annahme, dass solche Namen im Sinne der Warenzeichen- und Markenschutzgesetzgebung als frei zu betrachten wären und daher von jedermann benutzt werden dürften.

Verlag: Südwestdeutscher Verlag für Hochschulschriften Aktiengesellschaft & Co. KG
Dudweiler Landstr. 99, 66123 Saarbrücken, Deutschland
Telefon +49 681 37 20 271-1, Telefax +49 681 37 20 271-0, Email: info@svh-verlag.de
Zugl.: Hohenheim, Universität, Diss.,2008

Herstellung in Deutschland:
Schaltungsdienst Lange o.H.G., Berlin
Books on Demand GmbH, Norderstedt
Reha GmbH, Saarbrücken
Amazon Distribution GmbH, Leipzig
ISBN: 978-3-8381-0401-0

Imprint (only for USA, GB)
Bibliographic information published by the Deutsche Nationalbibliothek: The Deutsche Nationalbibliothek lists this publication in the Deutsche Nationalbibliografie; detailed bibliographic data are available in the Internet at http://dnb.d-nb.de.
Any brand names and product names mentioned in this book are subject to trademark, brand or patent protection and are trademarks or registered trademarks of their respective holders. The use of brand names, product names, common names, trade names, product descriptions etc. even without a particular marking in this works is in no way to be construed to mean that such names may be regarded as unrestricted in respect of trademark and brand protection legislation and could thus be used by anyone.

Publisher:
Südwestdeutscher Verlag für Hochschulschriften Aktiengesellschaft & Co. KG
Dudweiler Landstr. 99, 66123 Saarbrücken, Germany
Phone +49 681 37 20 271-1, Fax +49 681 37 20 271-0, Email: info@svh-verlag.de

Copyright © 2009 by the author and Südwestdeutscher Verlag für Hochschulschriften Aktiengesellschaft & Co. KG and licensors
All rights reserved. Saarbrücken 2009

Printed in the U.S.A.
Printed in the U.K. by (see last page)
ISBN: 978-3-8381-0401-0

Contents

1 INTRODUCTION 1

1.1 THE STARTING POINT 1
1.2 THE RESEARCH OBJECTIVE 2
1.3 A BRIEF OUTLINE OF THIS STUDY 2
1.4 DESCRIPTION OF THE STUDY REGION 4
    1.4.1 Natural conditions 5
    1.4.2 Regional economic situation 6
    1.4.3 Agricultural situation 6

2 SOIL SCIENCE ASPECTS 8

2.1 SOIL FUNCTIONS 8
2.2 A DEFINITION OF SOIL DEGRADATION AND ITS DIFFERENT ASPECTS 9
2.3 CURRENT CONDITIONS OF SOIL DEGRADATION 11
2.4 DEFINITION OF SOIL CONSERVATION 13

3 POLICY APPROACHES FOR SOIL CONSERVATION 15

3.1 INTERNATIONAL EFFORTS 15
    Excursus: a comparison between North America and Europe 17
3.2 NATIONAL LEGAL APPROACHES FOR SOIL CONSERVATION IN THE EXAMPLE OF THE GERMAN ENVIRONMENTAL POLICY 18
    3.2.1 General objectives of environmental policy 18
    3.2.2 Principles in the German environmental policy 19
    3.2.3 Juridical Instruments of environmental law in Germany 20
3.3 CURRENT INSTRUMENTS FOR SOIL CONSERVATION IN GERMANY 22
    3.3.1 The German soil protection act 22
    3.3.2 The role of agriculture in the Soil Conservation Act and other regulations 23
    3.3.3 Other political instruments for soil conservation in agriculture 25
    3.3.4 Criticism on existing policy approaches 27

4 THE ECONOMIC BACKGROUND FOR SOIL CONSERVATION 29

4.1 NATURAL RESOURCES AND WELFARE ECONOMICS 29
4.2 ECONOMIC PROPERTIES OF SOIL AS A NATURAL RESOURCE 31
    4.2.1 The degree of publicity – Soil as a private or public good 31
    4.2.2 Externalities 32
    4.2.3 Uncertainty and risk 33
    4.2.4 Irreversibility 37
    4.2.5 Substitution 37
4.3 CONCLUSION: MARKET FAILURE AND THE NEED FOR GOVERNMENTAL INTERVENTION . 38
4.4 POLICY INSTRUMENTS FOR MANAGING A NATURAL RESOURCE 39
    4.4.1 Policy instrument categories 39
    4.4.2 Achieving the social optimum through internalisation 40
    4.4.3 Spatial targeting and regulation areas 44
4.5 METHODOLOGICAL APPROACHES FOR THE ECONOMIC ANALYSIS OF SOIL CONSERVATION POLICIES 47
    4.5.1 Cost-Benefit-Analysis 47
    4.5.2 Cost-Effectiveness-Analysis based on a Safe minimum Standard 50
4.6 THE APPLICATION OF CEA WITH A SAFE MINIMUM STANDARD 53
4.7 THE ANALYSED IMPLEMENTATION OPTIONS 54
    4.7.1 Definition of instruments and measures 54

I

	4.7.2	Instruments	54
	4.7.3	Measures	56
5	**SOIL EROSION RISK ASSESSMENT OF SITES AND CROPPING PRACTICES THE EFFECTIVENESS OF SOIL CONSERVATION MEASURES**		**57**
	5.1	THE POTENTIAL EROSION RISK OF THE AGRICULTURAL LAND	57
	5.1.1	Erosion models for estimating soil erosion risk	57
	5.1.2	The soil erosion risk model in this study - an adapted USLE approach	59
	5.1.3	Erosion risk assessment of site conditions	60
	5.2	EROSION RISK ASSESSMENT OF CROPPING PRACTICES WITH A FUZZY-LOGIC MODEL	65
	5.3	COMBINATION OF EROSIVITY AND ERODIBILITY VALUES	68
	5.4	THE EFFECTS OF EROSION RISK THRESHOLDS IN SOIL CONSERVATION PROGRAMMES	70
	5.4.1	Threshold options for soil conservation programmes	71
	5.4.2	Applying erosion thresholds as a eligibility criteria	72
6	**A BIO-ECONOMIC MODEL FOR THE ANALYSIS OF POLICY INSTRUMENTS AND ON-FARM MEASURES**		**75**
	6.1	REVIEW OF ECONOMIC MODELS IN THE CONTEXT OF SOIL CONSERVATION	75
	6.1.1	Different ways of analysis	75
	6.1.2	Available modelling approaches	75
	6.1.3	An overview of policy relevant studies	76
	6.2	THE CHOSEN MODEL SYSTEM MODAM	78
	6.3	A CHOICE BETWEEN PROGRAMMING MODELS	83
	6.4	MODELLING THE AGRICULTURAL SECTOR OF THE STUDY REGION	85
	6.4.1	Costs	85
	6.4.2	Land data	86
	6.4.3	Labour supply	87
	6.4.4	Policy conditions	87
	6.4.5	Crop production	89
	6.4.6	Soil erosion	92
	6.4.7	Crop rotations	92
	6.4.8	Quotas	93
	6.4.9	Livestock systems	93
7	**RESULTS OF THE ECONOMIC AND ECOLOGICAL EVALUATION OF SOIL CONSERVATION POLICIES**		**98**
	7.1	FINDING RELEVANT OPTIONS FOR POLICY ANALYSIS	98
	7.2	BRIEF DESCRIPTION OF THE ANALYSED POLICY OPTIONS	100
	7.3	INDICATORS	102
	7.4	STATUS QUO SCENARIO: AGENDA 2000	103
	7.5	BASIC SCENARIO CAP2013 (CAP REFORM WITH DECOUPLED PAYMENTS)	106
	7.6	SOCIAL PLANNER SCENARIO	109
	7.6.1	Scenario description and trade-off curve	109
	7.6.2	Calculation of a shadow price for soil erosion	113
	7.6.3	Results of the benchmark scenario	114
	7.7	POLICY SCENARIO RESULTS	115
	7.7.1	Untargeted Incentive Scheme	115
	7.7.2	Targeted Incentive Scheme	118
	7.7.3	Targeted crop restrictions	119
	7.8	DISCUSSION OF THE MODELLING RESULTS	121
	7.8.1	Overall comparison of scenarios	121
	7.8.2	Analysis of erosion levels on high erodible soil types	125
	7.8.3	Thresholds results	127

7.8.4	Spatial analysis of erosion rates under different policy options	128
7.8.5	Changes in livestock production and labour needs	131
7.9	CONCLUSIONS	132
7.9.1	Policy options	132
7.9.2	The modelling system	133

8 TRANSACTION COSTS, PROPERTY RIGHTS AND SOIL CONSERVATION .. 135

8.1	BACKGROUND	135
8.2	THE SCOPE OF ANALYSIS	135
8.3	A BRIEF OVERVIEW ON NEW INSTITUTIONAL ECONOMICS	136
8.3.1	Transaction costs	136
8.3.2	Property Rights	136
8.3.3	An overview of the basic theories	138
8.4	TRANSACTION COSTS IN THE CONTEXT OF SOIL CONSERVATION PROGRAMMES	139
8.4.1	Transaction costs in policy evaluation	139
8.4.2	Boundary issues	139
8.4.3	Stakeholders and transaction costs	140
8.4.4	Forms of transaction costs	141
8.4.5	Recent attempts in measuring transaction costs of environmental policies	144
8.4.6	Suitable reference values for soil conservation policies	145
8.4.7	Attributes of an environmental good and transaction costs	146
8.4.8	Research on transaction costs in environmental policies	147
8.5	QUALITATIVE ANALYSIS OF TRANSACTION COSTS OF SOIL CONSERVATION POLICIES	148
8.5.1	Boundary issues	149
8.5.2	Agents involved in the policy making process	149
8.5.3	Forms of transaction costs	151
8.5.4	Attributes of participants and of the environmental good	153
8.6	CONCLUSIONS	153

9 DISCUSSION 155

9.1	THE THEORETICAL FRAMEWORK	155
9.2	THE BIO-ECONOMIC MODELLING APPROACH	155
9.3	THE RELEVANCE OF TRANSACTION COSTS	157
9.4	APPROPRIATE INSTRUMENTS FOR SOIL CONSERVATION	158

10 SUMMARY 160

11 ZUSAMMENFASSUNG 162

12 ACKNOWLEDGEMENTS 165

13 REFERENCES 167

List of Figures

Figure 1: Map of North-Eastern Brandenburg including the administrative district Uckermark and the study region "Prenzlau-West 4
Figure 2: Land use map of the study region Prenzlau-West 5
Figure 3: The influence of natural characteristics, socio-economic factors and land use on soil degradation 11
Figure 4: Optimal pollution tax t*, marginal external costs (MEC) and marginal net private benefits (MNPB) 41
Figure 5: Social optimum achieved through negotiations 42
Figure 6: Administrative and erosion risk units over two districts in Brandenburg 46
Figure 7: Universal Soil Loss Equation: Explanation and classification of factors 59
Figure 8: Erosion risk map of the study region grouped according to soil erosion risk categories 61
Figure 9: Aggregation commands used in GIS 62
Figure 10: Aggregation methods for soil erosion risk information using different statistical functions 63
Figure 11: Schematic representation of the combination of erosivity (C-factor) and erodibility (soil properties) 68
Figure 12: Potential soil erosion rates of crops with comparable tillage system (plough, standard practice) on sites with the highest erosion risk 69
Figure 13: Mean of soil erosion risk per crop including minimum and maximum value (n= numbers of cropping practices per crop and soil category type) 70
Figure 14: Percentage of possible eligible land for soil conservation programmes depending on the threshold levels 73
Figure 15: Data types used in the MODAM farm model 79
Figure 16: Three level, integrated economic and environmental analysis of agricultural land use systems with MODAM 80
Figure 17: Description of cropping practices in the MODAM database 90
Figure 18: Example layout of the LP-erosion module 92
Figure 19: Data describing a livestock system in MODAM 94
Figure 20: Gross margin (€/ha) and total erosion (tons) in model region - comparison of the Agenda 2000 and the CAP2013 scenario with decoupled payments 107
Figure 21: Share of high erosion crops in the Agenda 2000 and the CAP2013 reform scenario 108
Figure 22: Trade-off curve between total erosion levels and total gross margin in region based on a parameterized model run with increasing limitation on the erosion level 110
Figure 23: Share of reduced tillage with increasing levels of soil conservation in the model region (social planner scenario) 110
Figure 24: Share of set aside with increasing levels of soil conservation in the model region (social planner scenario) 111
Figure 25: Share of high erosive crops on high and low erosion risk field types on selected levels of erosion in the social planner scenario 112
Figure 26: Marginal costs per ton of reduced erosion in the region; based on the shadow price per ton of reduced total erosion at each step of total erosion restriction 114
Figure 27: Effect of the payment level for reduced tillage on the area under the conservation scheme and the total erosion in the region 116
Figure 28: Total soil erosion in the model region under different policy options 122
Figure 29: Costs of different policy options 123
Figure 30: Cost-effectiveness of different policy options based on total costs 124
Figure 31: Partial cost-effectiveness of policy options based on budget costs and on-farm costs 125

Figure 32: Erosion risk of crops on high erodible field types for different scenarios 126
Figure 33: Average soil erosion risk for the region under the CAP2013 conditions 129
Figure 34: Average soil erosion risk for the region under untargeted incentive conditions . 129
Figure 35: Average soil erosion risk for the region under targeted incentive conditions 130
Figure 36: Average soil erosion risk for the region under row crop restrictions................... 130
Figure 37: Boundary issues related to transaction costs 140

List of Tables

Table 1: Land use types in the study region Prenzlau-West and for the district Uckermark .. 5
Table 2: Different aspects of soil functions 8
Table 3: Human-induced Soil Degradation for the World, expressed in million hectares ... 12
Table 4: Soil conservation programmes co-financed by the EU directive 1257/1999 in the German federal states in 2003 27
Table 5: Components of the "Total Economic Value" of soil resources 30
Table 6: Categories of the site specific erosion risk for water erosion based on the ABAG-USLE assessment with a standard C-factor (0.11) 64
Table 7: Erosion risk–soil quality types as a combination of soil quality class and erosion risk category 64
Table 8: Possible combinations of statistical grid aggregation and threshold values 64
Table 9: Combinations of cropping practices implemented in the farm model 67
Table 10: Crop types implemented in the farm model 67
Table 11: Potential soil erosion risk for sugar beets depending on the soil erosion risk category 69
Table 12: Soil quality index classes used in the model and the respective tolerable soil erosion by water 72
Table 13: Overview for different threshold levels for soil erosion and references 73
Table 14: Area with erosion above threshold depending on grid size, aggregation method and threshold levels 74
Table 15: LP modules used in the model 86
Table 16: Surface distribution of different site qualities and erosion classes in the model region 87
Table 17: Subsidy levels in Agenda 2000 and the CAP-reform of July 2003 88
Table 18: Crop shares on arable land of study region in 2001 based on aggregated project data 89
Table 19: Crop yields for winter wheat depending on soil productivity class 91
Table 20: Example: Variables for the activity "winter barley" in the LP-matrix 91
Table 21: Crop rotation restrictions based on Good Technical Practice 93
Table 22: LP-coefficients for a dairy cow in the LP-model 95
Table 23: Livestock numbers for the district of the model region 95
Table 24: Animal groups in a dairy cow system of the model region 96
Table 25: Animal groups in the bull fattening system of the model region 96
Table 26: Hog production system in the model region 97
Table 27: Description and abbreviations used for the analysed scenarios 101
Table 28: Analysed indicators in each scenario 102
Table 29: Modelled and actual (2001) crop shares in percent on arable land 103
Table 30: Average erosion risk based on actual and modelled data on highly erodible field type 105
Table 31: Indicator values for the Agenda 2000 scenario 105
Table 32: Average and maximum soil erosion risks for crops on soil category 50_6 in the Agenda 2000 scenario 106
Table 33: Indicator values for CAP2013 scenario 106
Table 34: Average and maximum soil erosion risks for crops on soil category with highest erosion risk and best soil quality in the CAP2013 scenario 107
Table 35: Comparison of crop shares between Agenda 2000 and CAP2013 in percent of total area 108
Table 36: Livestock production in Agenda 2000 and CAP2013 scenario 109
Table 37: Crop shares within the study region for selected levels of erosion in the social planner scenario 111

List of Tables

Table 38:	Indicator values of the social planner scenario compared to the basic scenario (CAP2013)	114
Table 39:	Share of reduced tillage in the untargeted incentive scenario compared to the CAP2013-scenario	116
Table 40:	Indicator values of the untargeted incentive scenario compared to the CAP2013-scenario	117
Table 41:	Changes of crop shares in the untargeted incentive scenario compared to the CAP2013 scenario	117
Table 42:	Indicator values for the targeted incentive scenario compared to the CAP2013-scenario	118
Table 43:	Crop shares in the targeted incentive scenario compared to the CAP2013-scenario	119
Table 44:	Indicator values for the row crop restriction scenario with restricted cultivation of row crops on highly erodible field types compared to the CAP2013-scenario	120
Table 45:	Crop shares for the row crop restriction scenario (restricted cultivation of row crops on highly erodible field types) compared to the CAP2013-scenario	120
Table 46:	Overview of indicator values of the different policy scenarios	121
Table 47:	Share of reduced tillage in the different scenarios	122
Table 48:	Overview of crop shares under different policy options	123
Table 49:	Crop shares on highly erosive field type with good soil quality	127
Table 50:	Soil erosion risks of crops in scenarios compared to the threshold values for soil erosion risk	127
Table 51:	Animal numbers in the scenarios	131
Table 52:	Labour demand in the scenarios	131
Table 53:	Functions and examples of agents involved in the implementation of a soil conservation policy	141
Table 54:	Typology of transaction costs associated with public policies and parties incurring costs	141
Table 55:	Categories of transactional costs incurred in the implementation of voluntary schemes based on compensated management agreements and cost incidence	143
Table 56:	Appropriate instruments depending on the variability of the environmental good and producer/Production type	146
Table 57:	Policy approaches and administrative costs	147
Table 58:	Estimation of transaction costs types for the policy options in this study	151
Table 59:	Qualitative grading of the analysed policy options using transaction costs categories	152

Abbreviations

ABAG	Universal Soil Loss Equation (dt. *Allgemeine Bodenabtragsgleichung*)
BC	Budget costs
BMBF	German Federal Research Ministry
BMP	Best management practices
BMU	Federal Ministry for the Environment
CAP	Common Agricultural Policy
CBA	Cost-benefit-analysis
CEA	Cost-effectiveness-analysis
CES	Constant elasticity of substitution
CRP	Conservation Reserve Program
dt.	German
DüV	Fertilisation regulation (dt. *Düngeverordnung*)
EIA	Environmental Impact Assessment
EPA	US Environmental Protection Agency
EPIC	Erosion Productivity Impact Calculator Model
ESA	Environmentally Sensitive Areas
EU	European Union
FAO	World Food and Agriculture Organisation
GIS	Geographical Information System
GLADA	Global Assessment of Land Degradation and Improvement
GLASOD	Global Assessment of Human-induced Soil Degradation
GM	Gross margin
GRANO	Approaches for a sustainable agricultural production: Application for North Eastern Germany")
h	hour
ha	hectares
IACS	Integrated Administration and Control System (dt. *Integriertes Verwaltungs- und Kontrollsystem(INVEKOS)*).
IBSRAM	International Board for Soil Research and Management
ISCO	International Soil Conservation Organisation
ISRIC	International Soil Reference and Information Centre, Wageningen, The Netherlands
ITC	Institutional transaction costs
IUCN	International Union for Conservation of Nature and Natural Resources
IUSS	International Union of Soil Science
IWMI	International Water Management Institute, Sri Lanka
km	kilometre
LP	Linear Programming
M	million (10^6)
m	metre
MEC	Marginal external costs
MNPB	Marginal net private benefits
MODAM	Multiple Objective Decision Support Tool for Agro-Ecosystem Management
NIE	New Institutional Economics
OECD	Organization for Economic Cooperation and Development
PAV	Guidelines for the execution of Good Technical Practice in pest management (dt. *Pflanzenschutzverordnung*)

Abbreviations

PMP	Positive mathematical programming
SMS	Safe minimum standard of conservation
SOTER	Soil and Terrain Digital Database
SRU	German Advisory Council on the Environment (dt. *Rat von Sachverständigen für Umweltfragen*)
SSSI	Sites of Special Scientific Interest
t	metric tons
TC	Transaction Costs
TEV	Total Economic Value
TSE	Tolerable soil erosion by water
UNEP	United Nations Environment Programme
USLE	Universal Soil Loss Equation
VERMOST	Vergleichsmethode Standort
WASWC	World Association of Soil and Water Conservation
WBGU	The German Advisory Council on Global Change - Wissenschaftlicher Beirat der Bundesregierung Globale Umweltveränderungen
WEPP	Water Erosion Prediction Project
WOCAT	World overview of Conservation approaches and technologies
WTA	Willingness to accept
WTP	Willingness to pay
ZALF	Leibniz-Centre for Agricultural Landscape Research

1 Introduction

1.1 The starting point

This study was inspired by the implementation of the German soil conservation act, which passed the legislation process in 1998. This soil conservation act touches many aspects related to soil use, including agricultural use as well. However, a closer look at this act reveals that its impact on agriculture is rather limited. The effect is small because the law refers to non-controlled standards such as "good technical practice", which is a description of "proper" agricultural land use that is specified in a non-juridical sense. Soil scientists involved in the preparation of the law, proposed options that went further or were more elaborate, such as legal binding restrictions or subsidies for targeted areas (Wissenschaftlicher Beirat Bodenschutz beim BMU 2000). However, these suggestions were neither further specified nor implemented in the soil conservation act or in the respective enactments.[1]

These findings gave rise to the idea for a study that

1. analyses the effectiveness of policy options in the implementation of soil conservation measures, which go beyond the "best management practices" referred to in the German Soil Conservation Act and
2. describes the influence of property rights and transaction costs on the implementation process of policies.

This study focuses on the analysis of instruments and measures that can be used to promote more efficient soil conservation in Germany and gives suggestions for their implementation, so that the best practice and best policy options for soil conservation may be found[2].

Soils are seen as a non-renewable resource that needs protection from degradation. This definition is crucial for this study, since it sets the focus on the soil erosion process itself without the distinction of on- or off-site damages. Later conclusions in this study confirm that soils do need protection from degradation in order to fulfil the long term demands of society.

---

[1] Even to date, after the end of this study, agricultural soil use is not affected any further by this law.
[2] Parts of this study were achieved during an interdisciplinary, participative research project in Northeastern Germany (GRANO, "Approaches for a sustainable agricultural production: Application for North Eastern Germany") that was supported by the German Federal Research Ministry (BMBF). One of the aims of this project was to develop regionally adapted agro-environmental programmes in a co-operative manner with farmers, administrative agencies, environmentalists and other stakeholders through a round table ("Agri-Environment-Forum") (Arzt et al. 2003). Soil conservation issues and soil erosion processes were part of the agenda of this discussion process.

1.2 The research objective

The main objective of this study is to find decision support for the implementation options of soil conservation policies. Therefore, a combination of best practices (on-farm measures) and best policies (instruments) is needed for a more precise and efficient soil conservation and erosion control.

The aim of the study is to analyse and discuss the juridical, soil scientific, economic and agricultural aspects of soil conservation and to propose instrument-measure combinations for efficient soil conservation. Here, emphasis was given to both the resource and institutional economics of soil conservation. From an economic point of view, the scale on which an efficient promotion of erosion-avoiding measures is most adequate (best scale) and which instruments are best used for their promotion must be determined.

Therefore, it was important to develop an appropriate framework for the economic and environmental assessment of the implementation options. This framework was derived from the state of the art of economics and soil science.

The scientific work of this study depends on the consistent combination of soil science and economic theory for achieving an assessment tool that allows for the evaluation of different implementation options. The inclusion of new institutional economics allows for the consideration of transaction cost effects.

1.3 A brief outline of this study

The subject is approached in the following way: After the introduction and a short description of the study region in Chapter 1, Chapter 2 focuses on the relevant soil science aspects. Soil functions, definitions of soil degradation, current conditions of soils and further aspects of soil conservation are described in this chapter. The need for soil conservation is briefly highlighted by a short summary of soil conservation issues, and the related definitions of this aspect are given. Soil functions include more than just being an input for agricultural production. Soil serves also as a habitat for wildlife and plants, regulates ground water replenishment, and filters and metabolizes hazardous substances.

Chapter 3 contains approaches on soil conservation from the international to the national level. The instruments used for conserving and improving environmental conditions are discussed based on the general objectives and principles of the environmental policy in Germany. This is followed by an overview of the currently used instruments in soil conservation policies. Additionally, further proposals of possible policy changes are described.

Chapter 4 describes the economic theory that was applied in this research topic and sets the framework for the economic modelling approach. The role of property rights, public good characteristics of soils and the resulting externalities are discussed. The resulting market failure can be seen as the justification for a non-market coordination of soil use.

However, since the application of pure welfare economic instruments on the promotion of soil conservation shows some shortcomings in terms of data generation and applicability, a way to face the "coordination" problems of natural resources titled "A safe minimum standard of conservation" (Ciriacy-Wantrup 1963) is therefore illustrated. The resulting method is the cost-effectiveness analysis of the implementation options of soil conservation policies.

Chapter 5 describes the assessment of soil erosion risks based on both the natural conditions and the characteristics of the cropping practice. After an overview of the available soil erosion models, the chosen approach, which is an adapted Universal Soil Loss Equation (USLE) model, is described. Details are given for the underlying data used to describe the erosion risk of a region. The erosion risk of the cropping practices was analysed based on the application of a fuzzy-logic model. This procedure allowed for the provision of specific erosion risk values for standard and adjusted cropping practices.

Chapter 6 describes the design of the bio-economic model used in this study. After the description of the selection method for policy options, the structure of the model and the data used to describe the agricultural activities of a region are defined. The reason the applied model that was used to estimate the effects of implementation of policy instruments and measures for soil conservation was chosen is also outlined and justified here. With the use of a regional linear-programming model, the evaluation of the economic and ecological effects of agriculture was done using the example of a study region in Northeastern Germany.

Chapter 7 provides the results for a set of scenarios. First, a status quo solution based on the policy conditions of Agenda 2000 was calculated to show a starting point solution of the model and to check the model's plausibility against the original data. Based on these results a new scenario was designed that contained the main policy changes that were introduced by the 2003 CAP-reform. Then, a scenario that approached the issue from a social planner's viewpoint (one with the aim of reducing soil erosion) was analysed, with the assumption that the social planner has complete information on soil erosion rates for all cropping practices.

As main policy options, three policies were analysed based on the assumption of differing property rights concerning the right to cause soil degradation through soil erosion.

Based on a set of indicators, the cost-effectiveness of soil conservation policies was derived and evaluated. These results served as the basis for the further evaluation of policies in the following chapter.

Chapter 8 opens the scope to the aspects of institutional economics in soil conservation, discussing the influence of transaction costs and property rights on the success of soil conservation programmes. After an introduction of the theoretical foundations of this field of economics and their implications on policy making, a qualitative comparison of transaction costs and feasibility concerns on soil conservation programmes is presented. The results of the analysis highlight the relevance and the dimensions of transaction costs in terms of the cost of implementation, control and administration.

Chapter 9 draws some final conclusions on the theoretical framework, the bio-economic modelling approach, the relevance of transaction costs and, finally, the appropriate instruments for soil conservation based on the overall results of this study.

1.4 Description of the study region

The study region (total surface of more than 200 km²) is a mostly agriculturally used area situated in the administrative district "Uckermark", which is part of the federal state Brandenburg, 100 km north east of Berlin (see Figure 1).

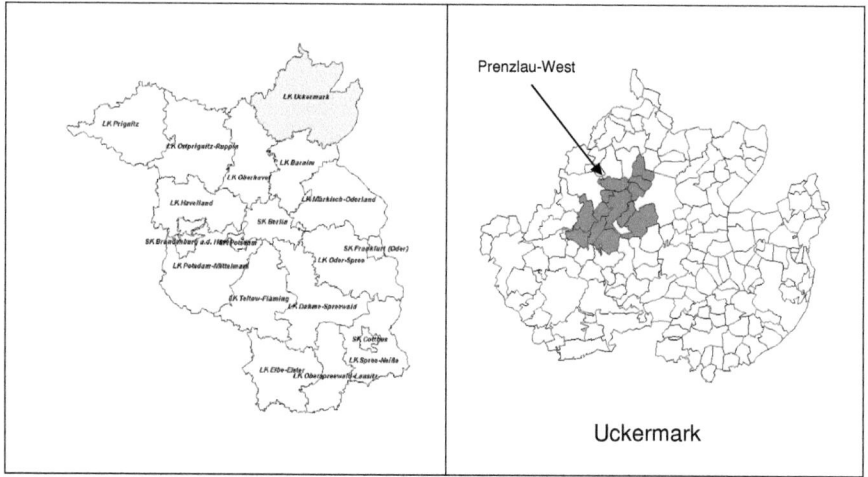

Source: Regiograph; own presentation

Figure 1: Map of North-Eastern Brandenburg including the administrative district Uckermark and the study region "Prenzlau-West"

The Uckermark is famous for its hilly landscape, which was shaped mainly by the ice ages (morainic landscape). In the southern parts of the Uckermark, the sandy soils of the outwash plains that resulted from the melt waters of the ice-age glaciers can be found. Environmentally, the selected study region "Prenzlau-West" is typical for the district, with arable farms, less structural elements and an open landscape to the North, while the Southern part shows more structural elements, forests and mixed farms. The share of land use types for the study region "Prenzlau-West" and the district Uckermark is shown on Table 1.

Chapter 1 - Introduction

Table 1: Land use types in the study region Prenzlau-West and for the district Uckermark

Area type	unit	Prenzlau-West	Uckermark
Total area of the region	km²	208	3058
Arable land	%	59.9	50.2
Pasture	%	11.2	9.0
Forests	%	16.6	22.3
Lakes, Rivers	%	2.7	5.1
Special habitats (§ 32)	%	7.3	7.4
Others (settlements, infrastructure)	%	2.3	6.0

Source: Landesumweltamt Brandenburg (LUA) 2002; Landesbetrieb für Datenverarbeitung und Statistik 2006

Figure 2 shows a land use map of the study region based on data from the biotope mapping in Brandenburg (Landesumweltamt Brandenburg (LUA) 2002).

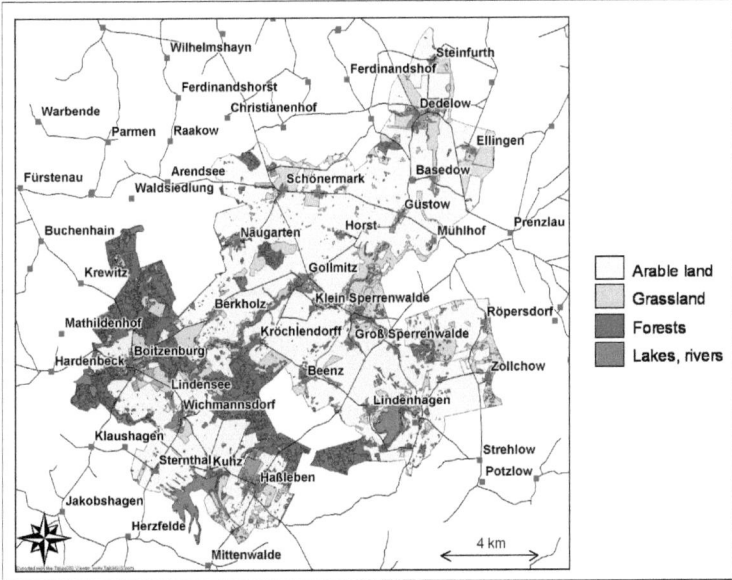

Source: Landesumweltamt Brandenburg (LUA) 2002

Figure 2: Land use map of the study region Prenzlau-West; based on biotope mapping data

This region was chosen because it is a "representative" region that covers the characteristics of a region in North-Eastern Germany in terms of heterogeneity of the landscape, land use and erosivity, with the advantage of reduced data needs.

1.4.1 Natural conditions

The Uckermark region has a yearly average temperature of 8.4 °C with an average precipitation of 486 mm per year (ZALF Müncheberg 2006). Another climatic characteristic often referred to by local farmers is the lack of rain during the late spring months (April, May), the months most important for plant growth.

The soils of Uckermark are roughly divided into more loamy type soils in the north, which provide relatively good conditions for agriculture, and lighter, sandy soils in the south (GIS-Data based on Schmidt and Diemann (1981)).

The water supply conditions of the soils seem to be non-problematic. It appears that potholes (dt. *Soelle*), a characteristic element of the North-eastern German landscape, do create some problems but only locally in agricultural land use. These relicts of the ice age are drainless water bodies in many fields, with changing water levels that affect the crops grown around them.

The entire region has a high number of lakes and small rivers. Although the landscape in northern Uckermark possesses fewer elements of scenic landscapes than the southern part, some features could still be found (e.g. potholes, hedges, old tree lined avenues, small woods along the fields).

1.4.2 Regional economic situation

The metropolitan, urbanized region of Berlin is potentially an appropriate sales market for the regional products of the region. However, the Uckermark region has not yet developed a suitable supply for this demand. North of the region lies the even less populated state of Mecklenburg-Western Pomerania that offers only a few possibilities as a market for agricultural products. Poland, the eastern neighbour of the region is seen economically more as a threat to the local economy than an opportunity. Agriculture however, is still an important economic factor in this region (Regionomica 2006).

The unemployment rate is around 25 % in the Uckermark district. Job opportunities are rather low. The landscape offers some points of tourist attractions, however the economic importance of tourism could be further developed (Regionomica 2006).

1.4.3 Agricultural situation

About two thirds of the district is used for agriculture (Landesumweltamt Brandenburg (LUA) 2002). The more fertile northern parts are usually worked by crop farms, while the share of livestock farms increases towards the south as a result of less fertile soils and fen pasture lands (Arzt et al. 2000; Arzt et al. 2002).

The most important cultivated crops are cereals, oil seeds and sugar beets. The production of renewable energy and industrial raw materials particularly the cultivation of non-food rapeseed reached a high level under the EU-set aside regulations (Landkreis Uckermark 2006; Ministerium für Ländliche Entwicklung 2006). Depending on the availability of premia for energy crops, this level might stay constant or even increase further. Animal husbandry comprised of cattle and pigs fell dramatically in numbers since the reunification of Germany (Bork et al. 1995).

In general, non agricultural income sources (e.g. tourism) play a smaller role in the local farms in comparison to the western part of Germany (Regionomica 2006).

Large scale, not individually owned farm enterprises (e.g. corporations, cooperatives) are still utilising approximately 60 % of the agriculturally used surface. Individual firms manage only 20 % of the agriculturally used surfaces (private companies manage the remaining 20 %) (Ministerium für Ländliche Entwicklung 2006).

2 Soil science aspects

2.1 Soil Functions

The awareness of soil functions is usually centered on the different aspects of production. However, there are other functions that are of high relevance even though they are not traded on markets. With increasing scarcity of non-degraded soils, their management is more and more subject to the ways of regulation, (e.g. landscape planning and land use jurisdiction).

By highlighting the different functions of soil, the need for soil protection can be underlined. The German Advisory Council on Global Change (Wissenschaftlicher Beirat der Bundesregierung Globale Umweltveränderungen (WBGU) 1994) classifies the functions of soils concerning their importance to plants, animals, micro-organisms and mankind as well as for the balance of energy, water and matter as follows (Table 2):

Table 2: Different aspects of soil functions

Function	German term	Example
Habitat	*Lebensraum*	Habitat for animals and plants
Regulation	*Regelung*	Regulation of thermic, hydrologic, physical and chemical processes
Use function	*Nutzung*	Soil itself can be used for the production of building materials (e.g. bricks, tiles etc.)
Production function	*Produktion*	Production asset in agricultural production
Spatial location function	*Träger*	Soil carries buildings, streets etc
Information function	*Information*	Soil contains information about pre(-historical) events
Cultural function	*Kultur*	Soils are a part of the cultural heritage of a country or landscape (e.g. Loess regions)

Source: Wissenschaftlicher Beirat der Bundesregierung Globale Umweltveränderungen (WBGU) 1994

Hannam and Boer (2002) provided a more detailed summary of soil functions. Their three main functional groups are natural functions, cultural functions and land use functions, with further subgroups (see following box).

Both definitions contain functions related to production, (land) use or spatial location of any construction and development. These functions are mostly used to generate income. In contrast, parts of the natural functions and the cultural functions do not usually generate income, but are, nevertheless, important for the functioning and survival of any society. The differing valuation of these functions is one of the causes of the mismanagement of a natural resource (see chapter 4.4.2).

Degraded soils cannot or can only partly fulfill the aforementioned functions. Therefore, the importance of viable soils is not only a focus in the production function for food and fibre (world food problem), but also a crucial condition for maintaining global biodiversity and for preserving the cultural heritage of societies. Non-degraded soil is the basic condition for globally sustainable development.

Box: Categories of soil functions

Soil functions

Natural functions

- Soil is the basis of life and living space for humans, animals, plants and micro-organisms.
- Soil is a fundamental element of nature and landscape
- Soil is part of the ecological balance, particularly with its water and nutrient cycles
- Soils provide a filtering, buffering and transformation activity, between the atmosphere, the ground water, and the plant cover, protecting the environment and especially humans through the protection of the food chain and the drinking water reserves
- Soils are used for agriculture and forestry to produce biomass
- Soils are biological habitats and gene reserves, much larger in quantity and in quality than all the above-ground biomasses

Cultural functions

- Soils are a geogenic and cultural heritage, which form an essential part of the landscape in which humans live, and
- which conceal paleontological and archaeological information of high value for the understanding of the history of earth and humankind

Land use functions

- Soils serve as a spatial base for technical, industrial and socio-economic structures and their development
- Soils are used as a source of raw materials
- Soils are a location for agriculture, including pastures and forestry

Source: Hannam and Boer 2002

2.2 *A definition of soil degradation and its different aspects*

The proper functioning of soils is threatened by soil degradation. The definition of soil degradation given by the German Advisory Council on Global Change (WBGU 1994) describes *"anthropogenic soil degradation as permanent and irreversible structural, functional changes in soils or their complete loss being caused by human induced physical, chemical or biotical stresses that exceed the recovering capacity of the soil systems"* (Translated by the author*)*. Schachtschabel et al. (1992) defined soil degradation (in this case soil erosion) as "a natural process on many sites on earth but globally, it is aggravated or even caused by the human use of soil" (translated by the author).

Bridges et al. (2001) listed a group of intrinsic factors for soil degradation, among them the influences of climate, terrain, vegetation and biodiversity and in particular, the soil biodiversity characteristics. However, the rate of human induced soil degradation is determined by other causes. Bridges et al. (2001) described these forces as:

- The biophysical (land use and land management, including deforestation and tillage methods);
- Socio-economic (land tenure, marketing, institutions, income and human health); and
- Political forces (incentives, political ideology) that influence the soil degradation processes.

Oldeman et al. (1990) distinguished five causative activities of human induced soil degradation:

- **Deforestation and removal of the natural vegetation**; usually for the reclamation of land for agricultural purposes, commercial forestry, road construction, urban development etc.
- **Overgrazing**; both the effects of removal or destruction of vegetation and trampling by livestock cause damage to soil cover, which increases the risk of water and wind erosion. Compaction can also be increased by trampling.
- **Agricultural activities** can cause soil degradation through improper management practices such as insufficient or excessive use of fertilizers, shortening of the fallow period, poor quality irrigation water, absence of anti-erosion measures, and inappropriate use of heavy machinery.
- **Overexploitation of vegetation for domestic use** is seen when vegetation is used for fuel wood, fencing, etc. Even though the vegetation is not completely removed, the remaining vegetation does not provide sufficient protection against soil erosion.
- **(Bio-)industrial activities** lead to the chemical and physical pollution of soils, causing soil degradation processes.

For further descriptions of degrading processes, see Hannam and Boer (2002).

The natural condition of soil can be more or less easily changed by the fore-mentioned factors and the rate at which it degrades depends on the influence of the human activities. Therefore, the rate of degradation is firstly the result of the natural conditions of the whole soil system, which is hardly changeable, but secondly the result of human land use. Land use affects soils and is the result of management decisions that are themselves dependent on a framework of socio-economic and political conditions (see Figure 3).

Economic and social conditions create new incentives and driving forces for changes in land use that can result in higher rates of soil degradation. Short-term incentives such as famine or non-sustainable, fast growing economies can increase soil degradation to a level that goes far beyond any recovery rate. Certain types of agricultural land uses are considered among the most important human induced driving forces of soil degradation (e.g. soil erosion) (Boardman et al. 2003). Therefore, finding ways to influence land use can be a very successful option for reducing soil degradation.

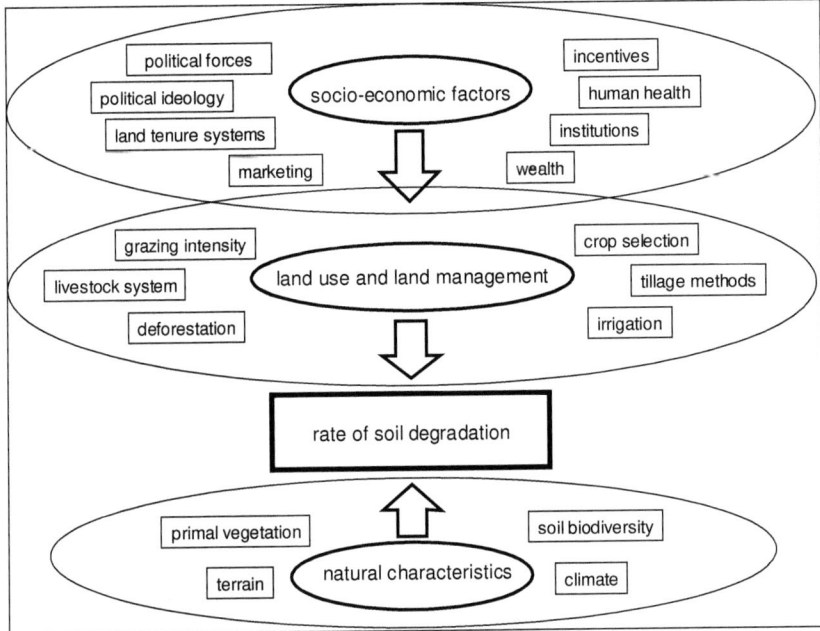

Source: own presentation; based on Hannam and Boer 2002

Figure 3: The influence of natural characteristics, socio-economic factors and land use on soil degradation

So far, it can be stated that soil degradation itself is like "a natural process on many sites on earth, which is globally aggravated or even caused by the human use of soil" (Schachtschabel et al. 1992). It only becomes a perceived problem when it exceeds levels that are no longer sustainable and jeopardizes the development and viability of a society. Nevertheless, degradation processes do increase the risk of irreversible losses of soil functions for the future generations.

2.3 Current conditions of soil degradation

Soils and their related viable functions are globally threatened by degradation through desertification, overuse, hazardous wastes and soil erosion by wind and rainfall. The fundamental survey in the "Global Assessment of Human-induced Soil Degradation" (GLASOD), which is still the only globally realised study on soil degradation showed that soil degradation is a threat to almost all countries in this world (Oldeman et al. 1990). According to this study, 1,964 M Hectares of land, representing 15 percent of the world's land surface, show signs of degradation. Water erosion is the most important factor for soil degradation, causing 56 percent of the degraded soils worldwide, followed by wind erosion with 28 percent. Table 3 gives an overview of the types of soil degradation and the surface being affected by it globally.

Table 3: Human-induced Soil Degradation for the World, expressed in million hectares

Type of degradation	Degradation classes (million hectares)					Percent
	Light	Moderate	Strong	Extreme	Total	
Loss of Topsoil	301.2	454.5	161.2	3.8	920.3	
Terrain Deformation	42.0	72.2	56.0	2.8	173.3	
WATER	**343.2**	**526.7**	**217.2**	**6.6**	**1093.7**	**55.6**
Loss of Topsoil	230.5	213.5	9.4	0.9	454.2	
Terrain Deformation	38.1	30.0	14.4		82.5	
Overblowing		10.1	0.5	1.0	11.6	
WIND	**268.6**	**253.6**	**24.3**	**1.9**	**548.3**	**27.9**
Loss of nutrients	52.4	63.1	19.8		135.3	
Salinization	34.8	20.4	20.3	0.8	76.3	
Pollution	4.1	17.1	0.5		21.8	
Acidification	1.7	2.7	1.3		5.7	
CHEMICAL	**93.0**	**103.3**	**41.9**	**0.8**	**239.1**	**12.2**
Compaction	34.8	22.1	11.3		68.2	
Waterlogging	6.0	3.7	0.8		10.5	
Subsidence organic soils	3.4	1.0	0.2		4.6	
PHYSICAL	**44.2**	**26.8**	**12.3**		**83.3**	**4.2**
TOTAL	**749.0**	**910.5**	**295.7**	**9.3**	**1964.4**	
Total Percent	38.1	46.4	15.1	0.5		100

Source: Oldeman et al. 1990

The rather out-dated data of GLASOD have led to the projection of a newer global approach "Global Assessment of Land Degradation and Improvement" (GLADA), but the project is still in the stage of development, so no data is yet available[3] (ISRIC - World Soil Information 2006).

A report on the environmental indicators of agriculture published by the OECD (2001) analyses soil erosion using two main indicators: risk of soil erosion by water and wind. This report stated that the condition of soil quality has gotten better in the past years in some countries, ever since the adoption of conservation and no tillage systems, which have led to a reduction in the run-off values. Nevertheless, for certain OECD countries, more than 10 per cent of their agricultural land fall within the risk class of high/severe risk (OECD 2001).

According to the OECD report, other aspects that decrease the quality of soil are acidification and sodification of soils, salinisation of soils, soil compaction, reduction in soil fertility and the increase in chemical and heavy metal pollution of soils.

As stated in the same report, "water erosion is causing considerable damage to soil fertility and ecological functioning in East Germany [*Note: where this study focuses empirically*], but the German Soil Protection Act (see Federal Ministry for the Environment 2003a) is beginning to address the problem"(OECD 2001). One of the objectives of this study is to investigate whether the German Soil Protection Act is really addressing this problem.

---

[3] Last update of website: 5/14/2007

The report "Soil erosion risk in Europe" (Grimm et al. 2002) found high erosion risks in the Mediterranean region due to the climatic conditions of long dry periods followed by heavy bursts of erosive rainfall that fall on steep slopes with fragile soils. The Northern parts of Europe are less prone to such high erosion rates due to climatic differences, but regions with hilly landscapes such as the example region of this study can have considerable rates when inappropriate cropping practices are applied.

In the German federal state of Brandenburg, where the example region of this study is situated, medium to high risk water erosion is estimated for 6 percent of its agricultural surface (Matzdorf and Piorr 2003) when the spatial comparison method VERMOST is used (Deumlich et al. 1997). This method does not provide information on the amount of erosion in the region, but rather a way to compare local proneness to water erosion. The mid-term review report of the Brandenburg agri-environmental measures (Matzdorf et al. 2003); also Appendix Map M2-5, p. 4) showed a concentration of medium to high erosion risks in the North-Eastern part of Brandenburg, underlining the heterogeneity of the water erosion problem. An extreme number was given by Frielinghaus et al. (1997), who reported erosion rates of 170t/ha from single events of strong rainfalls during periods of low soil coverage.

As shown in Table 3, soil degradation takes on many forms. In this study, **soil erosion by water as one form of soil degradation** will be the main topic because of the area affected in the test region and the amount of soil that can be lost through a single event of heavy rainfall.

2.4 Definition of soil conservation

Definitions of soil conservation "which tend to be technocentric" (Hannam and Boer 2002) are abundant and mostly similar. A typical definition is:

"The prevention, mitigation or control of soil erosion and degradation through the application to land of cultural, vegetative, structural and land management measures, either singly or in combination, which enable stability and productivity to be maintained for future generations (Houghton and Charman 1986; cited from Hannam and Boer 2002)."

Usually soil conservation is understood as the implementation of agricultural measures on the field level (e.g. contour ploughing, reduced tillage, terracing). However, soil conservation comprises also steps at the political, economic and juridical levels, given the fact that the main driving forces are created on these levels (Boardman et al. 2003; Bridges et al. 2001; Hannam and Boer 2002).

The International Board for Soil Research and Management (IBSRAM) (1997) defines sustainable land management as "land management systems that combine technologies, policies and activities aimed at integrating socio-economic principles with environmental concerns to satisfy the five pillars of sustainable land management." These pillars are (IBSRAM) 1997):

- maintain or enhance production,
- reduce the level of production risk,
- protect the potential of natural resources and prevent the degradation of soil and water quality,
- be economically viable and
- achieve social acceptability.

Morgan (1981; cited from Hannam and Boer 2002) defined the aim of soil conservation as obtaining the maximum sustained level of production from a given area of land whilst maintaining soil loss below a threshold level which, he said, theoretically permits the natural rate of soil formation to keep pace with the rate of soil erosion.

This definition contains two specific aims that can be used in an economic analysis. One is to find a *long term maximum level of soil use* that is not threatened by over-exploitation and nutrient mining. The other contains the idea of a *threshold rate of soil use or depletion* that can be adjusted to the natural rate of soil formation.

How these concepts are transferred into and used in economic theory will be taken up in chapter 4.

3 Policy Approaches for Soil Conservation

3.1 International efforts

The protection of soils is of great importance throughout the world. The number and quality of valuable soils are becoming more and more scarce and degraded by soil erosion and uncontrolled land consumption (Wissenschaftlicher Beirat der Bundesregierung Globale Umweltveränderungen (WBGU) 1994). The "Brundtland"-report stressed the need for sustainable development that include the prevention of further soil degradation as well (The World Commission on Environment and Development 1987).

In North America, soil conservation research and extension services have played a dominant role since the 1930s after the effects of non-sustainable soil use in the prairies was experienced (Furtan and Hosseini 2003).

Hurni (2003) described the history of soil conservation attempts at the international level starting in the 1970s. Numerous workshops and meetings had been held addressing soil erosion and land degradation at the international level. Organisations like FAO, WASWC, IBSRAM (now IWMI), IUSS and ISRIC[4] were involved. Some projects for assessing the global effects of soil degradation such as "Global Assessment of Human-induced Soil Degradation" (GLASOD, continued through GLADA) (ISRIC - World Soil Information 2006), which showed that soil degradation is a threat to this resource in almost all the countries of the world (Oldeman et al. 1990), were started. Another project that assesses soil degradation is SOTER (Soil and Terrain Digital Database), a joint program organized by ISRIC, FAO and UNEP[5]. The WOCAT project (World Overview of Conservation Approaches and Technologies) was started in 1992 as a result of the 7th ISCO[6] conference to improve the information on tools and approaches for soil conservation on the global level. For further international projects addressing soil conservation see (Hurni 2003).

Another approach by the scientific community to try to set soil degradation on the international political agenda was the agreement to a concept of a convention for the protection of soils, which requests national governments to place soil conservation on a legal basis (Held and Kümmerer 1997). The influence of such a convention is still discussed among politicians and scientists, but the need for more efficient soil protection measures cannot be denied (Hurni 2003).

---

[4] FAO (World Food and Agriculture Organisation), WASWC (World Association of Soil and Water Conservation), IBSRAM (International Board for Soil Research and Management), IWMI (International Water Management Institute, Sri Lanka), IUSS (International Union of Soil Science), ISRIC (International Soil Reference and Information Centre, Wageningen, The Netherlands)
[5] (UNEP) United Nations Environment Programme
[6] (ISCO) International Soil Conservation Organisation

Hannam and Boer (2002) stated in a report by the IUCN[7] Environmental Law Centre the poor recognition of the need for internationally more binding soil protection, and emphasised the importance of an international legal framework to help avoid further global soil degradation. They postulated the need for ecological soil conservation standards or norms for ensuring sustainability. Although erosion is a problem with a very local physical impact, the effects of eroded and degraded soils are of international relevance. This means, actions are needed on a regional, national and international level. On the regional level, measures and programmes that are adapted to local conditions need to be developed to provide effective soil protection. On the national level, a soil law should set the legal fundaments for these regional measures. However, both national and regional laws will not be implemented without an international treaty or convention, which sets equal binding standards for all signing states and avoids competition for the least restricting legislation.

Since soil degradation as a problem shows its effects more in the long run, politicians will be tempted to base their decisions on short term considerations to be re-elected (Buchanan et al. 1980). A sustainable soil conservation policy that shows its effects only in the long run, but which places restrictions on farmers in the short run while decreasing their competitiveness on the world market, will not be very popular. However, when embedded in an international agreement, a soil conservation law is more likely to mediate the objectives of farmers, politicians and soil scientists. These agreements vary among non-binding instruments (e.g. the *Rio Declaration on Environment and Development* 1992), binding instruments (e.g. *United Nations Convention to Combat Desertification* 1994) or non-government initiatives such as the proposal for the *"Convention on Sustainable Use of Soils"* (Held and Kümmerer 1997).

Looking at the huge number of organisations involved in soil conservation world-wide, there does not seem to be a lack of knowledge and research dedicated to the conservation of soils world-wide. Hurni (2003) stated that a gap still exists between research and stakeholders like farmers and farm extension agencies that has to be filled with local agendas, so that technological knowledge and the awareness for more sustainable land management can be transferred from research to land users.

It is questionable whether a mere global approach on either the scientific or political level will improve the management of soils. Nevertheless, these projects can set a framework for national, regional and even local approaches to improve sustainable land use by setting binding goals for action (Khan 1993). Environmental law is an essential component for setting and implementing global, regional, and national policy on the environment and development.

---

[7] (IUCN) International Union for Conservation of Nature and Natural Resources

On the EU-level, soil conservation had been the subject of a non-binding instrument since 1972, when the Council of Europe adopted a *European Soil Charter* as a part of the attempt to stop the steady deterioration of land in Europe (Hannam and Boer 2002).

In 2002, the European Commission published the communication "Towards a Thematic Strategy for Soil Protection" (Commission of the European Communities 2002). Since then, several EU-working groups have developed a legal framework that assures the protection of soils from their major threats. A wider overview on soil related policy activities on the EU-level is shown in Kraemer et al. (2006). The final objective of the strategy is a framework directive that was designed similar to the water framework directive.

This thematic strategy was passed by the European Commission in September 2006, instructing the member states to monitor soil conditions in their countries and to set-up measures to counteract harmful effects of soil degradation (Commission of the European Communities 2006). However, the first attempt to reach a political agreement on the draft directive for establishing a framework for the protection of soil did not attain a qualified majority in the European Council meeting of the Ministers for the Environment in 2007 (European Council 2008b).

Excursus: A comparison between North America and Europe

North American soil conservation strategies (e.g. the Conservation Reserve Program (CRP)) have been more targeted and focused on the prevention of off-site effects as compared to the European approaches (Plankl 1999). European soil conservation efforts are focused mostly on the prevention of erosion in general without the consideration of where the damage occurs. So far, only a few European environmental programmes have targeted soil conservation measures to specific erosion prone areas (e.g. North Rhine-Westphalia (Hartmann et al. 2006)). To some extent, the European legislation relies mostly on the proper management of resources by land users, e.g. as pointed out in the soil protection act (Federal Ministry for the Environment 2003a). Financial incentives for the adoption of conservation practices are usually available to all farmers of a region for reasons of fairness (Hartmann et al. 2006); (Huylenbroeck and Whitby 1999). Furthermore, soil erosion problems in Northern and Central Europe never reached such extremes seen in the North-American Prairies (Masutti 2004).

The differences may lay in the history of soil erosion problems: farmers were faced with huge problems in the North-American Prairies, when soil erosion by wind and water was threatening both farmers directly and society indirectly through the loss of future food resources (Furtan and Hosseini 2003). The problem started as an individual problem that turned into a problem for the society, which justified the grounding of intensive research and extension (Furtan and Hosseini 2003; Masutti 2004). After years of farm extension on soil conservation in North America (Furtan and Hosseini 2003; Popp et al. 2002), society continues to concentrate on the off-site effects, while the on-site effects are left to the decision of the farmer, who decides on the more or less sustainable use of his property (Crosson 1984).

3.2 National legal approaches for soil conservation in the example of the German environmental policy

The use of soils is usually regulated at the national level. In most countries, the aspects of soil conservation are usually covered under specific laws that are not aimed directly at soil conservation. The use of soil or land is rather regulated so that specific aims may be achieved (e.g. nature conservation) or property rights among potential users may be regulated. Hannam and Boer (2002) listed potential soil use regulating laws as follows:

- land administration,
- pastoral land management,
- maintaining of biodiversity,
- conservation of native vegetation,
- forest law,
- environmental protection,
- environmental planning and assessment.

In order to give a clearer picture of the situation in Germany, the German legal approaches on soil conservation are described in the following. The general objectives and principles of the German environmental policy are first highlighted, since they serve as a guideline for environmental policy making. Then, the instruments of environmental policies in Germany are shown. The current instruments for soil conservation in Germany are then illustrated based on this. The chapter closes with a brief description of the shortcomings of this legal system in relation to soil conservation, argued from all the different sides.

3.2.1 General objectives of environmental policy

In the Federal Republic of Germany, the environmental policy is based on three main objectives. Firstly declared in a governmental environmental programme in 1971, the environmental policy comprises all measures needed to meet the following three objectives (Storm 1988):

1. ensuring an environment for people, which meets their needs for health and a humane existence,
2. conserving soil, water, air, fauna and flora from harmful effects of human impact,
3. eliminating damage and harm from human impacts.

By referring to human activity and interest in all three objectives, it is underlined that environmental conservation is seen from an anthropocentric point of view, which sets conservation in relation to the maintaining and improving of human welfare in the broadest sense. Nevertheless, conflicts between concurring objectives like agriculture and nature conservation will always be common and will need to be solved through the evaluation of the individual case (Storm 1988).

3.2.2 Principles in the German environmental policy

In order to achieve the objectives of the German environmental policy, three main principles will need to be applied (Freshfields 2003; Storm 1988):

3.2.2.1 The principle of precaution (Vorsorgeprinzip)

This principle aims at avoiding any damage before it even occurs through preventive measures and regulations. By applying this principle sustainability can be ensured, as a pure reactive approach will only regulate liability after the damage has occurred and precious resources irreversibly destroyed. This principle is of high importance in the soil conservation legislature, because soil degradation can only be reversed with huge efforts.

3.2.2.2 The polluter pays principle (Verursacherprinzip)

The polluter pays principle is used as a means to allocate the costs of environmental effects of any human action, but it is not yet an allocation of the total liability. Any person, who endangers, pollutes or causes damage to the environment is held liable for the costs of avoiding the damage and the clean-up. Exceptions are made when economic distortions occur or when the polluter cannot be tracked down. In those cases, the public would be responsible for paying for the damage. For soil conservation this principle is a crucial point: most of the EU-countries hardly apply this principle when it comes to off-site damages from agricultural fields (Boardman et al. 2003). On the contrary, this principle is often applied when soils are contaminated by industrial uses.

3.2.2.3 The co-operation principle (Kooperationsprinzip)

This principle sets the directions of how the state and society should interact in environmental policy, as long as existing laws do not interfere. The involvement of stakeholders should assure better decision quality in environmental issues and prevent future claims by persons affected by the planned activities. Participation processes are common in any big transaction or investment with a possible impact on the environment. In soil conservation, approaches such as incentives for less damaging agricultural practices could be grouped into this category. Even though the polluter pays principle could be applied, the existing set of property rights allows farmers to use their soils according to good technical practice, which includes a certain amount of erosion (see chapter 7.9). Therefore, farmers and the community agreed in a type of voluntary process to negotiate on a new level of soil erosion.

An additional aspect in policy making is seen in the option of establishing laws that are valid without restrictions or laws that are based on subsidiarity, which means that the law is only applicable, if no other law already regulates the specific case.

3.2.3 Juridical Instruments of environmental law in Germany

Schmidt and Müller (1992) listed the instruments available for the management of natural resources in Germany. These are:

- Planning instruments
- Instruments of regulatory law (*Ordnungsrecht*) such as preventive and repressive prohibitions
- Environmental impact assessment (*Umweltverträglichkeitsprüfung*).
- Fiscal instruments such as subsidies, taxes and duties

These instruments are discussed in the context of their applicability to improvement in soil conservation.

3.2.3.1 Planning instruments

Most of public or private investments are accompanied by an obligatory planning process (dt. *Fachplanung*) by public and private planning agencies, (e.g. house building or construction of new processing plants in industry). During this process the possible conflicts with all relevant laws are investigated before any development can begin. Only when this process proves that there is no violation of any law or regulation, can the actual construction begin. The instrument of urban land use planning (dt. *Bauleitplanung*) must also consider the effects on the environment.

Furthermore, environmental conservation is promoted by a more general spatial planning that is based on the regional planning act (dt. *Raumordnungsgesetz*). This law sets the guidelines for zoning in urban and rural areas based on economic developing objectives and priority setting for the construction of public infrastructure projects like schools, roads etc. Among other things, the goal of environmental issues needs to be taken into account in order to protect valuable sites from destruction or deterioration.

The relevance of planning instruments in soil conservation in agriculture is limited. Some importance is seen in the zoning of certain areas to allow only certain crops or cropping practices.

3.2.3.2 Instruments of regulatory law

Administrative rules (dt. *Ordungsrecht*) allow public agencies to influence the environmental effects of most economic and private actions. There is a long list of instruments that are usually applied:

1. registration and report obligations (dt. *Anmelde- und Anzeigepflichten*)
2. disclosure obligations (dt. *Auskunftspflichten*)
3. security obligations (dt. *Sicherungspflichten*)
4. preventive and repressive prohibitions (dt. *Präventive und repressive Verbote*)
5. injunctions (dt. *Verfügungen*)

Agriculture is subject to many of such instruments: pest management (PAV 2006), construction of agricultural buildings, manure and fertilizer use (DüV 2006), which are regulated by such guidelines. So far, soil use has not been directly addressed by such instruments. However, soil use is indirectly affected by them. One possible option of using such an instrument is to implement certain soil conserving practices through the law, or to legally make farmers report off-site damages, as is already outlined in the soil protection act (Federal Ministry for the Environment 2003a; Federal Ministry for the Environment 2003b).

3.2.3.3 Environmental impact assessment

Certain projects may even call for a special Environmental Impact Assessment (EIA) (dt. *Umweltverträglichkeitsprüfung*) with specific regulations on how the impact on the environment is to be evaluated (e.g. power stations, chemical plants, railway tracks or airports). This regulation is based on a EU-guideline that demands the application of this instrument in all member countries within a three year period (art. 12, European Council 2003). The applicant has to provide information on the possible impacts of the planned project that will be evaluated by the authority. Furthermore, a public hearing is required in this procedure. Finally, a comprehensive report of the environmental effects has to be published. Soil conservation on agricultural fields however is too decentralized to be subject to such an assessment.

3.2.3.4 Fiscal instruments

German law also allows the use of economic instruments for influencing environmental behaviour. Depending on the objective, a subsidy, tax, or a duty is applied for, the instruments can be structured as follows (Schmidt and Müller 1992; Storm 1988):

Environmental taxes and subsidies: Taxes can be used to decrease the use of scarce resources or to internalize negative external effects, as the increasing price of production will decrease the use of the natural resource. Subsidies are seen as a negative tax offered by the government to enforce economic activities that would not occur or at least not at the wanted level without this financial transfer. While taxes are more or less uncommon in the European agri-environmental policy making, subsidies are widespread for the promotion of agricultural activities demanded by society. Examples are tax reductions for cars with lower exhaust gas pollution, or differing taxes on hydrocarbon fuels in accordance with their pollution effects (e.g. unleaded fuel).

Environmental duties: Duties are used to either steer the use of resources in a more sustainable way (guidance duties) (dt. *Lenkungsabgaben*), or as a compensation for damage done to the environment (compensation duties) (dt. *Ausgleichsabgaben*). Duties can be seen as a fee that has to be paid for the use of a resource.

Both these instruments (taxes and duties) create the same economic effect because both will increase the price of the resource and will at the end decrease the consumption of the resource. The difference between taxes and duties is based more on legal reasons and the purpose the money is used for. According to the legal definition, taxes have to create at least to a certain degree public income and do not entitle the taxpayer to claim for any governmental action in exchange. By contrast, duties are seen as a compensation to a legal body for efforts in waste clean-ups or damages to the environment.

Both instruments are of increasing importance in the political process, leaving the methods for damage reduction to the knowledge of consumers and producers trying to avoid the costs of pollution. However, governmental regulation and control still plays a major role in environmental conservation (Freshfields 2003).

Standard taxes for soil degradation have not been used in soil conservation. However, subsidies for soil conserving practices are used commonly in agri-environmental programmes (Hartmann et al. 2006). Furthermore, the concept of ambient taxes for non-point source pollution such as from soil erosion had been discussed in the economic literature (Segerson 1988). This tax is imposed on all possible polluters of a given area, irrespective of the individual pollution. Economic theory shows that such a tax can reduce pollution even if producers with high pollution levels are not identified.

3.3 Current instruments for soil conservation in Germany

3.3.1 The German soil protection act

Soil use in Germany is regulated through a multitude of laws and guidelines from different categories as follow:

- the fertilisation regulation (*Düngeverordnung*) (DüV 2006), which controls the use of manure and synthetic fertilisers on agricultural plots,
- the guidelines for the execution of Good Technical Practice in pest management (*Pflanzenschutzverordnung*) (PAV 2006)
- the general principles of good technical practice of agricultural land use, which have not been legally specified except under the above regulations DüV and PAV and
- the Federal Soil Protection Act (Federal Ministry for the Environment 2003a), which was implemented in 1998.

Soil and legal scientists had stated the need for soil protection based on an independent law before the implementation of the soil protection act, as the pressure on soils from human activities had created demands for preventive measures in soil protection (Wissenschaftlicher Beirat der Bundesregierung Globale Umweltveränderungen (WBGU) 1994). It was argued that this goal could be better achieved with the help of a specific soil conservation law than with a patchwork of singular laws aimed at specific aspects of land use (Landel et al. 1998).

Finally, after years of discussion at scientific and political levels, a federal German soil protection act (Federal Ministry for the Environment 2003a) was brought on its way and finally passed in the federal legislation (*Bundestag*) on March 17th 1998. The case for and the aim of such a law was nevertheless doubted throughout the discussion process. Critics of the specific soil conservation act stated that soil protection is still better achieved through several, specific laws that already exist, whereas advocates of the law demanded a more binding frame law that emphasises the importance of soil protection and conservation.

According to Landel et al. (1998), the legislators have two main objectives:
- the establishment of precaution aspects in the soil related legislation and
- the harmonisation of residual waste legislation.

The law passed encompasses elements of prevention, emergency procedures (*precaution principle*) and restoration/clean-up of polluted soils (*polluter pays principle*) (Landel et al. 1998), showing two of the aforementioned principles of German environmental legislation with elements of cooperation being less implemented (see Chapter 3.2.2). The law is based on subsidiarity, so other laws that already regulate the specific cases are given priority. This restriction had to be made, since there are already existing specific laws concerning soil (e.g. regulation of herbicide use or hazardous waste, see above). The law is accomplished through an ordinance that guides the administrator in the use of the law and sets threshold values for the contamination of soils (Federal Ministry for the Environment 2003b). Therefore, the soil protection act sets the frame for the instruments of regulatory law (see Chapter 3.2.3.2).

The soil protection law aims at the avoidance of any harmful deterioration of soils and regulates the legal effects of hazardous substance contamination. The law is not aimed at the restoration of soils to its almost natural state: the functions listed in the law include also the use function for production. It however sets limits on uncontrolled use of this natural resource.

3.3.2 The role of agriculture in the Soil Conservation Act and other regulations

A separate article of the law is dedicated to agriculture, describing how agricultural practices should be accomplished while considering preventive aspects of sustainability. The box below cites article 17, which regulates the agricultural use of soils.

The fundamental principles related to agriculture comprise (Landel et al. 1998):
- the maintenance and improvement of soil structure
- avoiding soil compression and run-off
- conservation of structure elements of the landscape

- conservation and improvement of the biological activity of the soil through rotational aspects
- conservation of organic matter (humus).

Box: German Federal Soil Conservation Act

Description of the German Federal Soil Conservation Act (Article 17)

Part Four
Agricultural Soil Use
Article 17
Good Agricultural Practice

1. In cases of agricultural soil use, the obligation to take precautions pursuant to Article 7 shall be fulfilled by good agricultural practice. In their advising, the competent agricultural advising bodies pursuant to federal state (*Länder*) law should impart the principles of good agricultural practice pursuant to paragraph (2).
2. The principles of good practice in agricultural soil use are the permanent protection of the soil's fertility and of the soil's functional capacity as a natural resource. In particular, the principles of good agricultural practice include:
    - in general, the soil shall be worked in a manner that is appropriate for the relevant site, taking weather conditions into account,
    - the soil structure shall be conserved or improved,
    - soil compaction shall be avoided as far as possible, especially by taking the relevant soil type and soil humidity into account, and by controlling the pressure exerted on the soil by equipment used for agricultural soil use,
    - soil erosion shall be avoided wherever possible, by means of site-adapted use, especially use that takes slope, water and wind conditions and the soil cover into account,
    - the predominantly natural structural elements of field parcels that are needed for soil conservation, especially hedges, field shrubbery and trees, field boundaries and terracing, shall be preserved,
    - the soil's biological activity shall be conserved or promoted by means of appropriate crop rotation and
    - the soil's humus content, as is typical for the site in question, shall be conserved, especially by means of adequate input of organic substances or of reduction of the intensity with which the soil is worked.
3. Obligations pursuant to Article 4 shall be fulfilled by means of compliance with the provisions mentioned in Article 3 (1); where these provisions contain no requirements for prevention of hazards and no such provisions result from the principles of good agricultural practice pursuant to paragraph (2), the other provisions of this Act shall apply.

Source: Federal Ministry for the Environment 2003a

The soil conservation act however, sets relatively small restrictions on agriculture, since sufficient soil degradation protection can be achieved through the "Good Technical Practice". However, the article does define such rules on a legal basis, which was not the case before the implementation of the soil protection act (see above box, part 2).

As mentioned before, agricultural soil use was already regulated under the fertilisation regulation (*Düngeverordnung*) (DüV 2006), the guidelines for the execution of Good Technical Practice in pest management (*Pflanzenschutzverordnung*) (PAV 2006), as well as the general principles of good technical practice of agricultural land use, which have not been legally specified.

As Landel et al. (1998) pointed out, the soil protection act does not add too much to those existing regulations. Additionally, the article is formulated in such a way that all other laws concerning soil use would be used primarily, even if the soil protection law would set higher restrictions or demands on sustainable and adequate soil use (see *subsidiarity*). The authors saw this as one of the weakest parts of this enactment (Landel et al. 1998).

Due to the federal system in Germany, the soil protection law had to be implemented at the federal state level. Some federal states had existing specific soil protection laws (e.g. Berlin, see Landel et al. 1998) that had to be revised according to the frame conditions of the new federal law. However, some federal states have not yet passed the soil protection law and its corresponding ordinances. In the case of Brandenburg, a federal state law on soil protection was still not passed in 2008, while other federal states had already implemented such acts. In Brandenburg, soil use is still regulated under other soil related laws, which underlines the low impact the federal soil protection act actually has.

3.3.3 Other political instruments for soil conservation in agriculture

3.3.3.1 Cross compliance

Since the implementation of the new CAP-reform (European Council 2006a; European Council 2006c), *Cross compliance* as a new instrument also affects agricultural soil use. Member states have to set up minimum standards for resource protection, nature conservation and animal welfare. Farmers have to meet these standards in order to receive EU direct decoupled farm payments. If farmers violate these rules of Cross Compliance, the EU-direct payments can be withdrawn. Cross compliance is in fact an application of the instruments of regulatory law (see Chapter 3.2.3.2).

For Germany, the following cross compliance rules affect soil conservation issues directly:

- **Erosion reduction** (DirektZahlVerpflV 2006) §2): no ploughing on 40 % of arable land between after harvest and February 15th unless a new crop is sown before December 1st. Local authorities can cancel this article if erosion risk is low for the region or if the weather conditions do not allow the application of this regulation. Human made terraces in agricultural plots may not be removed.

- **Conservation of soil organic matter** (DirektZahlVerpflV 2006) §3): on the farm level, a minimum number of three crops have to be grown, with a minimum share of 15 % per crop. If more than 3 crops are grown, smaller shares can be added up to reach the 15 % minimum. In the case where less than 3 crops are grown, farmers have to monitor soil organic matter by a balance through a scientifically approved method and prove values that are above thresholds specified in the regulation. Furthermore, the burning of stubble fields is forbidden.
- **Preservation of grassland shares within regions** (European Council 2006c): Member states have to show proof of a constant grassland share within regions to the European Commission. For Germany, a region comprises a federal state. If the share of grassland decreases under the specified limits, the federal state has to show proof of legal steps to stop such trends.

Other regulations such as the preservation of landscape structures (e.g. hedge rows) (DirektZahlVerpflV 2006) §5) have an indirect influence on soil conservation, e.g. by cutting slopes into shorter parts or by providing protection against wind erosion.

3.3.3.2 Agri-environmental measures

Member states of the EU have introduced agri-environmental programmes based on the Council Regulation (EC) No 1257/1999 (European Council 2006b) that also allow the support of soil conservation measures. Some federal states in Germany have introduced programmes that provide incentives for the adoption of soil conservation practices such as reduced tillage and direct seeding (see Table 4).

The amount of subsidy provided in these programmes varies between 42 €/ha and 117 €/ha. Some federal states support reduced tillage only for row crops with a higher erosion risk, others target these programmes on areas with a high erosion risk or support investment in special equipment (Brand-Saßen 2004). For a complete overview of agri-environmental programmes under the EU 1257/1999 regulation see Hartmann et al. (2006). Since 2007, the agri-environmental programmes had been based on the Council Regulation (EC) No 1698/2005 on support for rural development by the European Agricultural Fund for Rural Development (EAFRD) (European Council 2008a).

Table 4: Soil conservation programmes co-financed by the EU directive 1257/1999 in the German federal states in 2003

Federal state	Programme acronym	Programme objectives	Amount of subsidy (€/ha)
Bavaria	KULAP A	Reduced tillage in row crops with cover crops	100
Baden-Württemberg	MEKA II	Reduced tillage with cover crops	60
Brandenburg	KULAP	Cultivation of small seed legumes	310
Hesse	HEKUL	Reduced and no tillage with cover crops	60
Lower Saxony	NAU	Reduced and no tillage	72
	AFP	Investment in reduced tillage equipment	allowance up to 20 %
North Rhine-Westphalia	KULAP	- Reduced and no tillage for beets, corn, rapeseed, legumes and potatoes with cover crops as well as cereals	102
		- Conversion of arable land to grassland	306-715 (depending on soil quality)
		- Both parts of the programme are only available in targeted regions based on soil erosion risk	
Rhineland-Palatinate	FUL	Reduced tillage for corn and sugar beets	
		- no cover crops	46
		- with cover crops	117
Saarland	-	Reduced tillage	60
Saxony-Anhalt	-	Reduced tillage	42
Saxony	UL	Reduced tillage	25
		With cover crops	66
		Under sown crops	51
		Investment in Reduced tillage equipment	allowance up to 35 %
Schleswig-Holstein	-	Reduced tillage	60

Source: Brand-Saßen 2004; Hartmann et al. 2006

The efficiency of such programmes is still under discussion. In a comparison of the agri-environmental programmes in North Rhine-Westphalia and Rhineland-Palatinate with the example of model farms, Busenkell (2004) found that the soil conservation programmes (reduced tillage) either overcompensated the losses incurred by the programme, causing windfall gains; or on the other extreme, did not cover the costs through yield losses, which resulted in low adoption rates. For Brandenburg, it was demonstrated that soil conservation programmes showed little spatial focus on areas where soil erosion risk is elevated, resulting in the low efficiency of such programmes (Matzdorf et al. 2003).

3.3.4 Criticism on existing policy approaches

Despite taking part in the formulation of the soil conservation act, the German Scientific Advisory Council on Soil Conservation did outline their opinion that good technical practice does not guarantee sufficient prevention of soil erosion (Wissenschaftlicher Beirat Bodenschutz beim BMU 2000). The Advisory Council criticized the missing specification of minimum standards for the "good technical practice" that meant an ecological evaluation and estimation of the erosion effect of the appropriate production procedures was not possible. In comparison to examples from fertilisation regulation, or the guidelines for the execution of good technical practice in plant protection or the principles and recommendations for good technical practice in agricultural land use, it shows that the "aspects of soil protection are not yet sufficiently substantiated" especially in terms of more binding guidelines in support of soil conservation.

Both the Enquete-Kommission (1994) and the German Advisory Council for the Environment (Der Rat von Sachverständigen für Umweltfragen (SRU) 1996) considered the specifications for the "good technical practice" guidelines insufficient for guaranteeing sustainable soil management. They had suggested further, partly restrictive measures for the conservation of soils and proposed management restrictions, obligations and/or limitations for the cultivation of fields, land use limitations and retraction of certain measures. The Scientific Advisory Council for the Environment demanded as well for a more precise definition of "good technical practice" that contains more obligatory rules. Additionally it recommended a more purposeful promotion of erosion-avoiding measures, which cover the regional aspects of soil erosion as well, i.e. to consider particularly erosion-endangered fields or regions (Der Rat von Sachverständigen für Umweltfragen (SRU) 1996).

However, in a paper submitted by the scientific advisory board (Wissenschaftlicher Beirat Bodenschutz beim BMU 2000) neither the precise measures of soil protection were specified, describing their expected effect (*practical issues*), nor was the amount of the compensations mentioned. Beyond that, the conceptions are still very vague with regards to the political implementation or negotiability of the appropriate measures (*policy issues*). Furthermore, the question of how the assignment of subsidies can be bound more strongly to the prevailing erosion potential arose.

Overall, it can be stated that soil conservation in Germany is still lacking in further implementation, which comes either from more binding law applications or from increasing the efficiency of the incentive based soil conservation measures (through more adequate measures and/or spatial targeting of such programmes).

4 The economic background for soil conservation

While soil science demanded a rather absolute prevention of soil degradation processes (Frielinghaus et al. 1998), economists brought up the notion that any use of a resource has both positive and negative effects on different individuals, on different sites and at different times (Dabbert 1994; Hampicke 1991; Pearce and Turner 1990; Pimentel et al. 1995; Tisdell 1991).

Therefore, the economic question arose, as to whether, from a normative viewpoint (i.e. searching for an increase in total welfare of society), a socially optimal soil degradation rate should be derived, or, to find efficient ways of soil conservation based on a pragmatic, given threshold facing the empirical difficulties to reach a social welfare optimum.

In order to base the analysis on a theoretical framework, a brief overview of the economic characteristics of the decision problem is given in the following chapter. The available methods for solving such problems are additionally discussed. After that, an approach that meets the conditions and restrictions of the empirical decision problem best is derived based on a consistent economic background.

4.1 *Natural Resources and Welfare economics*

In resource economics, a welfare economics approach that is based on utility maximization is usually used to analyse and evaluate the management of natural resources (such as soil and soil conservation). This approach can be described as a sustainable management of a natural resource (Kooten 1993; Pearce 1993; Pearce and Turner 1990).

In general, the analysis compares solutions to a pareto-optimal state, where the utility of an individual cannot be further increased without the utility of another worsening. Any state that is not pareto-optimal implies welfare losses to society. In a less strict criterion for a social optimum, the individual with the increased utility is at least capable of compensating the individual whose utility had decreased, even when this compensation does not happen (Kaldor-Hicks-criterion) (Tisdell 1991). Welfare economics can be used in a theoretical or more conceptual way to analyse the rents generated by a policy.

Under perfect conditions with competition, perfect information and clearly defined property rights, markets achieve the social optimum through the coordination of "the invisible hand" (Smith 2005) of the market. Therefore, market-based prices are usually the most efficient instrument for the management of most resources; however, there are conditions that can lead to market failure with a sub-optimal provision of the specific good.

Crucial elements for market failure are (Hampicke 1991; Tisdell 1991):

1) public good characteristics,
2) the resulting presence of externalities,
3) unsustainable levels of resource use based on uncertainty and risk including the divergence of individual and social interest rates,
4) the irreversibility of the resource use,
5) the low possibility of substitution of the natural resource.

Other aspects are monopoly rights and common property that can lead to conditions where markets no longer provide the most efficient management of the resource. In such cases, governmental intervention may be indicated (Kula 1992; Pearce and Turner 1990; Tisdell 1991).

Governmental policies for resource management are therefore a means to more efficient use of resource from society's viewpoint (Cansier 1993; Hampicke 1991; Tisdell 1991; Weersink et al. 1998).

Given the deficiencies of using pure market values for the evaluation of natural resources, the "Total Economic Value" (TEV) of a natural resource was introduced as a construct to define and measure values for natural resources that go beyond the market prices (Pearce and Turner 1990; Turner et al. 2003). Pearce (1993) defined the total economic value of a natural resource as the sum of use values and non-use values (see Table 5). The use value consists of direct and indirect values.

Table 5: Components of the "Total Economic Value" of soil resources

Total Economic Value				
Use value			Non-use value	
Direct value	Indirect value	Option value	Bequest value	Existence value
Agriculture Forestry	Flood protection Filtering CO_2 sink	direct and indirect values for future generations	heritage for future generations	The value of mere existence, i.e. not related to any human use, but only as a part of nature

Source: modified, based on Pearce 1993

The direct value of soil expresses the production function for agriculture or forestry for producing direct income. Turner and Jones (1991, cited in Pearce 1993) expanded the TEV-approach by introducing another use value: the primary value that expresses the ecosystem functions of soil. The indirect value expresses soil ecosystem functions associated e.g. with flood protection, cleaning, buffering, storage, cooling as well as habitat functions. The option value corresponds to the direct and indirect value for future generations, which is the value to use to maintain options for future generations e.g. to grow food and fibre in a sufficient way but also to provide indirect functions for the future (see Chapter 4.2.4). Non-use values are defined as bequest and existence values. Human beings desire to bequeath sound soil resources to future generations including inter alia the cultural heritage function of soil that can be evaluated and expressed as a bequest value (Tisdell 1991). The existence value of soils describes society's appreciation of the mere existence of different soil types (intrinsic value) that goes beyond conventional utilitarian approaches (Pearce 1993). For a critical review on the different approaches on how to define use and non-use values, see Cicchetti and Wilde (1992). A recent overview of evaluation techniques can be found in Turner et al. (2003).

In the following, the characteristics of soil and soil use are analysed with the help of economic categories to find the appropriate framework for further analysis of soil conservation strategies.

4.2 Economic Properties of Soil as a Natural Resource

4.2.1 The degree of publicity – Soil as a private or public good

From an economic viewpoint, the optimal management of any resource is strongly dependent on whether it shows the characteristics of a private or public good (Tisdell 1991). Public goods show no rivalry in their usage and nobody can be excluded from their usage (Henrichsmeyer et al. 1991). In the case of public goods, prices cannot reflect the scarcity of the good, so the production of the good would be insufficient. The same is true for public "bads" such as environmental pollution. The malfunction of price signals can cause an "overproduction" of pollution (Hampicke 1991; Tisdell 1991).

At first glance, soil used by a farmer for agricultural production can be seen as a private good, for a farmer usually holds the production rights for a field. He or she has the right to use and change the good, others can be excluded from its usage, and there is rivalry for the use of the grounds[8].

However, the use of soil as a private good, being defined with absolute exclusion of other users, could harm other individuals through the run-off from a field. Obviously, there are effects from such "private land use" on other individuals caused by the run-off of soil matter washed on to other fields or into waterways that affect the water quality for people living downstream of the eroded fields (Crosson 1984).

Additionally, soil degradation affects the long-term fertility of the soil, which, under certain conditions, can be less important to the farmer than to society (Dabbert 1994). This can be seen as a non-rivalry effect of soil use. All individuals of a society as a whole would suffer from this loss without being able to exclude themselves: A non-rivalry condition versus long-term degradation. The public costs arising from soil degradation will most probably not influence the farmer's decision (Kiker and Lynne 1986). Therefore, the long-term fertility of soils can also be seen as a

---

[8] The possibility of commonly used soils such as grazing lands is neglected here.

public good, representing an option or bequest value for society (see Table 5). This will be further discussed in chapter 4.2.3.

4.2.2 Externalities

Public good characteristics of a resource often lead to externalities, since the non-rivalry and non-excludability would lead to an over-consumption or exhaustion of a resource (Tisdell 1991). In the case of soil degradation, the resulting effects would be soil use above sustainable levels. The example of soil use shows that the off-site effects of soil use would become a public good (or bad, respectively) (Barbier 1995). Externalities are defined as effects on other individuals that are not part of an economic transfer (Cansier 1993; Pearce and Turner 1990). Such externalities are effects of economic activities that change other individuals' welfare and this welfare change is not compensated (Pearce and Turner 1990), p. 61). The change of welfare is expressed in the external costs of an economic activity. The external costs of soil degradation can also be defined as their social costs, since these are not internalized into the producer's decision. Hence, these costs are burdened on society (i.e. other individuals). Given the differing social and private production costs of a good, a producer will use the resource to a higher extent as if he would also face the additional costs that society or other individuals have to bear (Tisdell 1991). In the example of soil use, the producer would choose a production level or technique that degrades soil more than what is optimal from society's perspective. In most cases, degraded soils will not only decrease the farmer's income (Evans 1995), but the external effects of degradation through soil erosion will also harm neighbours and society (Pimentel et al. 1995).

Soil conservation measures do not only affect the farmer's income as a private good. They can also be considered as a public good, since nobody can be excluded from the benefits of less damage. Many studies stressed that farmers' private costs are by far exceeded by the social costs of off-site damages from soil erosion (Boardman et al. 2003; Hansen et al. 2002; McConnell 1983; Commission of the European Communities 2006). The resulting effects can range from long term trends like decreasing soil fertility (which threatens food security), and increasing sediments in rivers and dams, to short term catastrophic events like muddy floods that cause considerable damages in residential areas (Boardman 1995; Boardman et al. 2003; Clark 1985; Evans 1996; Grimm et al. 2002; Hansen et al. 2002; Pretty et al. 2000).

Therefore, soil as a resource can show different characteristics: The resource itself is used as a private good but shows public good characteristics, which lead to externalities. As a result, it can be stated that soil degradation through soil erosion is to a large extent a public good (bad). Since the preferability of the market solution holds only true for private goods, the existence of public good properties of soil degradation through soil erosion can cause the sub-optimal use of the resource. Therefore, external intervention of the management of a natural resource is appropriate (Tisdell

1991). Furthermore, soil degradation threatens the long-term fertility of soils (Dabbert 1994). Thus, there is a value in the long-term fertility of soils, which is more important to society than to the individual. This adds a further argument for governmental action.

> **Excursus: The economic research on soil conservation in North America and Europe**
>
> North-American economic literature on soil conservation and soil erosion (for a short overview see (Furtan and Hosseini 2003) is much more profound in comparison to the European literature. Most articles follow the idea that soil erosion in general is not a problem to society as long as the erosion only harms the polluter/owner (Fox et al. 1995). Therefore, economic studies on the preferability of instruments for soil conservation take the location where the damage is occurring and where this damage can be avoided at the least cost more into account than trying to reduce erosion in general (Nakao and Sohngen 2000; Westra et al. 2002; Yang et al. 2003).
>
> European literature on the economics of soil erosion and soil conservation is still rather limited. Dabbert covered the issue in a general approach on the economics of soil fertility (Dabbert 1994), but publications in scientific journals are still rare (Jarosch and Zeddies 1991) and limited to the farm scale (Brand-Saßen 2004; Busenkell 2004; Meyer-Aurich 2005) or to certain regions (Boardman et al. 2003; Evans 1996; Pretty et al. 2000; Schuler and Kächele 2003).

4.2.3 Uncertainty and risk

An important assumption made in economics for the efficient functioning of markets as the optimal way of resource management is the existence of certainty regarding price signals. However, reliable information about future prices is almost impossible to predict, since the influence of time brings at least a range of probabilities of the possible outcomes (Hampicke 1991; Pearce and Turner 1990).

4.2.3.1 Definitions and relevance of uncertainty and risk

There is a difference in the definitions of risk and uncertainty (Ciriacy-Wantrup 1963; Pearce and Turner 1990). The distinction is that uncertainty is not measurable whereas risk can at least be described through expectation values (i.e. probability) (Bishop 1978).

Potential soil erosion can be influenced by the probability of heavy rain falls throughout the year and other climatic factors. However, there is uncertainty about many erosion related processes such as the replenishment rate of soils, since knowledge about these processes is still not very profound (Shortle and Miranowski 1987).

4.2.3.2 The divergence of individual and social interest rates

Interest rates are usually used to link and compare economic decisions from different time periods. In the context of resource conservation, interest rates play an important role in solving dynamic decision problems through being coordinated by the market or other economic instruments (e.g.

taxes). Individual and social interest rates are also distinguishable (see Dabbert 1994). The definition and their effects are given in the following:

Individual interest rates

Individual interest rates are a result of time preference, which serve as the compensation for the time span between activities and the flow of money. Time preference is a phenomenon that is seen in almost any economic activity. Economic literature states that individuals and society have differing time preferences (Pearce et al. 1990).

It is a widely accepted fact that individuals prefer the payment of a certain amount in the present rather than in the future. Pearce et al. (1990) gave some explanations for this behaviour.

The time preference of individuals is affected by:
1. the uncertainty of the future,
2. impatience,
3. the risk of dying, and
4. the decreasing net benefit of consumption.

Impatience is a very human characteristic. Therefore, an immediate payment is given a higher value than one in the future. In fact, it is a very rationale behaviour for individuals to prefer an earlier payment because there is always the risk of dying. Furthermore, future payments are affected with uncertainty, especially if the payment is scheduled far into the future. The source of payment could disappear, the bank could go bankrupt or the economy could deteriorate with a severe inflation of the currency. Furthermore, when considering the growth of the individual income, the marginal net benefit of a payment in the future would not be as high as its value in relation to the present income (Dabbert 1994).

Social interest rates

Social time preference originates from the opportunity costs of capital, for it is possible to derive income from the productivity of capital. The marginal return from the last used unit of capital is the social interest rate. Basically, the social interest rate is the aggregated interest rate of individuals that had been cleared of any external effects that could affect an individual's interest rate (Dabbert 1994).

The effect of an interest rate is to discount future benefits from the present value so that the preference for the two investment options may be compared. This is well accepted for the evaluation of private investments such as new machines or factories because the new investment has to compete with other opportunities. However, when the effects of the investment in public goods, and certainly the conservation of exhaustible natural resources are analysed, more difficulties are produced (Pearce and Turner 1990). The conservation or preservation of a natural resource for future generation is faced with a much longer time horizon than the usual investment in assets. The choice of a discount rate that is too high would decrease future benefits expressed in present values to almost zero and therefore allow for a greater degradation of soils. For that reason, some authors even suggested an interest rate equal to zero for the conservation of a natural resource, while others suggested rates below that of financial markets (Ciriacy-Wantrup 1963; Dabbert 1994; Hampicke 1991). However, given the close relation between production output and soil erosion, setting the discount rate at zero would only overvalue future benefits from soil use (and soil conservation) and possibly not allow any production at present for the sake of future benefits. As a further effect, any investment in other options (other than soil use) would then be more profitable.

In optimization or assessment approaches, the level of the interest rate used for discounting the future net benefits of public projects or policy programmes determines the most preferable option (Markandya and Pearce 1991; Tisdell 1991).

The above mentioned uncertainty regarding the actual amount of soil loss related to certain soil uses and the natural replenishment rate will widen the range for an appropriate choice of discount rates (Shortle and Miranowski 1987).

In the economic literature on soil erosion, the proper functioning of interest rates as a means that temporally interlinks soil use had been questioned. Even though it has been shown theoretically (mostly based on McConnell (1983), who abstracted from any off-site effects) that interest rates are able to coordinate inter-temporal soil use, empirical difficulties (Ervin and Mill 1985; Kiker and Lynne 1986) and conceptual questions (Hampicke 1991), lead to the conclusion that the choice of the discount rate is one of the main problems faced in the management of exhaustible natural resources (Pearce and Turner 1990).

The above aspects show how critical it is to rely on price signals for the sustainable management of soils; prices are affected by uncertainty, which can cause distortions in the proper reflection of scarcities. As a result, it can be stated that soil use is affected by uncertainty and risk, implicating distorted price signals, and therefore, soil resource management can be subject to market failure.

4.2.3.3 Uncertainty in economic soil erosion models

Most economic models on soil erosion assume a very rough linkage between production level and soil erosion. Usually, only the total output is correlated to the total amount of erosion (e.g. in models like McConnell (1983)) or a simple assumption of the erosivity of different cropping systems is made (Walker 1982).

The above mentioned model by McConnell (1983) describes the inter-temporal, economic effects of soil erosion. The aim of McConnell's model is to find out whether farmers have an incentive to use their soils in a sustainable way. Off-site effects were excluded in this model even though they were stated to cause the higher overall costs. The model is basically an optimization function of a capitalized profit stream over time that uses the soil both as a production asset and as a good with a future resale value. The level of soil loss is assumed to be correlated to the level of production output. Soil depth is an exhaustible resource that can be depleted within a certain time if production causes higher soil loss than replenishment by natural formation.

While high levels of erosion can decrease the resale value of the farm, they do also increase the profits during the production period. McConnell stated that it is economically rationale to accept a certain amount of soil degradation. Theoretically, assuming perfect functioning market mechanisms, a farmer's economic rationale would lead to the integration of soil conservation measures into private decision making through the matching of marginal costs of soil erosion with the marginal costs of soil conservation. According to McConnell (1983) farmers have in most institutional conditions an invested interest in soil conservation because they want to maximize the total income of their farms. Since farmers know that soil base affects the farm resale value, they are willing to apply appropriate conservation measures. McConnell's approach was strongly questioned by Kiker and Lynne (1986), who showed that market signals do not guarantee the proper conservation of soils according to societal needs.

Firstly, the information of how much soil is actually lost through a certain production system is still difficult for the farmer to estimate during the production process. In fact, even soil research is still not satisfied with the current state of soil erosion risk modelling. Therefore, soil loss is not a variable that a farmer could actually vary according to the prices he gets for his products. Secondly, the lack of an appropriate interest rate for coordinating soil use produces wrong signals for how much soil may be depleted (see Chapter 4.2.3.2).

The problem of modelling uncertainty in soil erosion economics was presented by Shortle and Miranowski (1987). Even though they underlined that it is analytically possible to identify the optimal level of erosion, they concluded that the empirical specification of such models causes problems that make the use of such optimization models questionable. These problems arose from estimating the soil losses related to cropping systems, soil types and spatial aspects. Therefore, it is not clear in most models as to what extent agricultural measures will contribute to soil conservation. The condition of soil can be described by soil depth, which is assumed to be a variable that can be influenced directly by the land users (e.g. in models based on McConnell (1983)) or can be related more directly to variables such as cropping systems. Such models are still lacking in enough variability in production methods and machinery use (Yang et al. 2003).

Factors such as sparse information on replenishment rates or the actual erosion rate from specific cropping systems make it impossible to manage the sustainable use of soils through pure market mechanisms. Full information is fundamental for the generation of precise price signals that lead to an optimum (Arrow et al. 1995; Grepperud 2000; Kiker and Lynne 1986). Ervin and Mill (1985) derived in an analytical model the finding that land prices could incorporate signals for the proper management of soils. However, their findings were based on an appropriate level of information of the relationship between production and soil erosion processes. However, the uncertainty in this relationship is the crucial aspect within this system, not the function that relates cause and effect (Shortle and Miranowski 1987; Tisdell 1991).

4.2.4 Irreversibility

Irreversibility is evident for the destruction of many exhaustible natural resources. Soil erosion processes are hardly reversible in spite of technical solutions that compensate for soil loss and degradation (e.g. fertilizers) (Dabbert 1994). Calculations of optimal soil run-off can be too optimistic, causing the long-term degradation of a valuable resource for the future nutrition of the world's population (Ciriacy-Wantrup 1963). Once a natural resource is completely exhausted, the time horizon for its rebuilding can be beyond human dimensions (The World Commission on Environment and Development 1987). Degradation of soils is more or less a one way process (Wissenschaftlicher Beirat der Bundesregierung Globale Umweltveränderungen (WBGU) 1994). It might be even economically profitable to exhaust soils beyond the sustainable limit in a certain period, even though the opposite would be preferable in the following period, but the chance to reverse this process would be zero then (Arrow et al. 1995; The International Board for Soil Research and Management (IBSRAM) 1997; Van den Born et al. 2000).

Furthermore, option values (see Table 5) that express individuals' value for keeping the option for future usage face difficulties in the way of how the value is expressed in monetary units. Therefore, the actual value of current use is overestimated (Tisdell 1991). Bequest values, expressing the value for future generations (Pearce 1993) are likely to be threatened by the irreversible effects of free-riding in the case of a public good (Tisdell 1991).

For the markets to function, these non-use values have to express the irreversibility of resource use. Given the public good characteristics of a resource, these price signals will not develop appropriately (Walsh et al. 1984). These effects combined with the aforementioned uncertainty regarding soil processes can cause a pure market solution to become an inefficient management option.

4.2.5 Substitution

From an economic viewpoint, a sustainable use of a resource is given, when the overall capital stock stays equal, i.e. the transfer from the natural resource capital to man-made capital generates a higher social return than when keeping its original state (Pearce and Warford 1993). The underlying assumption is that natural resource capital can be substituted with man-made capital. This poses two fundamental questions: First, is it possible at all to compare the values of natural resource capital with man made capital, and second, which values should be used to estimate the returns of a natural resource.

The first question is a question of definition. While some advocates of strong sustainability would argue against such comparisons (Batie 1989), environmental economists do not oppose the comparison, but concentrate rather on a clear comprising definition of the values to be compared (Pearce and Turner 1990; Pearce and Warford 1993) so as to find an answer to the second question. Table 5 shows the different types of values that are used to describe the total value of a resource.

The knowledge of existent values however does not lead to a clear answer for both questions. So, the possibility of substitution can either be argued by definition or the result of the substitution depends on factors such as discount rate chosen (see chapter 4.2.3.2) or the range of values introduced into the analysis (see above).

A more pragmatic answer to whether soil can be substituted at all can be given from a technical viewpoint: The resource soil and its long-run fertility can hardly be substituted or, with present means, can only be substituted at prohibitively high costs (Dabbert 1994).

The loss of soil fertility can be partly replaced by the use of fertilizers and other inputs, but this is only possible up to a certain degree and with declining marginal returns (Kiker and Lynne 1986). Moreover, the rate of substitution between natural fertility and technical inputs is not known. Soil degradation was accompanied by technical progress in the past that overlaid the negative effects of the decrease in natural fertility (Barbier 1995).

As a result, the limited possibility of substitution of soil through other resources and the uncertainty of possible costs stemming from an underestimation of indirect and non-use values decrease market prices' reflection of social preferable levels of scarcity. Therefore, soils should be used with an adequate level of caution that reflects the uncertainty und irreversibility of the degradation processes (Dabbert 1994).

4.3 Conclusion: Market failure and the need for governmental intervention

Based on the framework by Hampicke (1991), both soil use and soil conservation showed in all aspects at least some evidence that a pure market-based soil use management is questionable and will therefore not lead to a socially optimal solution. Market failure is a very likely result if the use of soils would be based only on price signals that are developed on markets. Therefore, other ways of resource management will have to be found.

Given the above mentioned arguments, it can be assumed that soil degradation in relation to agricultural production means:

- an externality with social costs exceeding the private costs,

- that these costs of the resource use are not optimally internalized given the public good character of the off site effects (market failure), and,
- that a social optimization procedure will face difficulties in selecting the appropriate interest rate for the discounting of the future costs and benefits (theoretical difficulties).

Due to the irreversibility of degradation processes and the limited possibilities of substitution, soils will not be conserved appropriately when only market signals are used for conservation decisions.

In referring to the normative viewpoint (i.e. to search for an increase in the total welfare of society), the net social benefit should be maximized instead of only private benefits. The Producers' internalization of external costs is one way of achieving this objective. In such cases, governmental intervention is indicated. Soil conservation policies are therefore ways to internalize such effects to achieve more efficient resource use from society's point of view.

However, market failure does not justify un-coordinated governmental intervention, as it might only lead to inefficient public expenditures. It is important to clearly analyse the subject and find cost-efficient solutions, so that market failure that turns into government failure can be avoided. Environmental economics provide instruments for both the internalisation of externalities and the management of public goods, and tools to analyse the efficiency of such governmental instruments. Furthermore, the consideration of transaction costs can lead to more efficient solutions within this process (see Chapter 7.9 and 8).

4.4 Policy instruments for managing a natural resource

4.4.1 Policy instrument categories

Based on the above conclusion (that the use of the pure market mechanisms shows deficits for managing the sustainable use of soils), policy instruments for resource management are required. Weersink (2002) compiled a general list of policy options for influencing farmers' behaviour and the resulting environmental performance: Moral suasion, command-and-control strategies and incentive-based strategies (see also Weersink et al. 1998; Weersink 2002). A list of instruments by Tisdell (1991) is further specified as:

1. taxes
2. subsidies
3. auctioning of rights to engage in externality-producing activities
4. state ownership and control of property
5. strengthening of property rights
6. internalisation of externalities by extension of ownership
7. fiat, prohibition or regulation
8. facilitation of private negotiation and agreement

9. provision of information.

Although some are not economic instruments, most of them can be evaluated using an economic framework. While taxes and subsidies influence the behaviour indirectly by changing the relative prices of in- or outputs (1-2), all property rights related instruments change or strengthen the institutional conditions around the use of a resource (3-6). This is also true for prohibitions and regulations (7). Additionally, the effect is similar to a tax, but without creating income for the government (Tisdell 1991). The provision of information as well as facilitation of negotiation processes are aimed at the reduction of transaction costs, which could be too high for the proper functioning of a market solution (8-9).

Another option for categorizing policies is the scale the policies are aimed at (Tisdell 1991): some policies address worldwide changes in resource consumption (greenhouse gases, desertification) (The World Commission on Environment and Development 1987), others focus on local or regional adjustments for a more sustainable resource use (i.e. the use of cost-benefit-analysis for development projects (Bishop 1978)), and a third group tries to tie global and local efforts together i.e. the Agenda 21 (UN Department of Economic and Social Affairs 2006).

4.4.2 Achieving the social optimum through internalisation

In the following, two theoretical approaches are shown by applying welfare economics analyses: the first is a tax-based solution based on the works of Pigou (1998), which explains the functioning of subsidies and the second is a negotiation approach following the argumentation by Coase (1960), which underlines the importance of property rights. Both approaches lead to a socially optimal allocation of the resource and the maximization of net social benefits. Therefore, most policy instruments are based to some extent on the ideas of these examples. Relative prices are either changed politically to induce a change in behaviour or the property rights are defined more clearly or actually changed so that change in the resource use is induced. Information instruments are not described here, since they only change the frame conditions of market situations that lead to conditions where markets themselves provide an efficient allocation of resources.

4.4.2.1 Pigou taxes

Given the problem of externalities (i.e. the off site effects of soil erosion) that decrease the welfare of society, Pigou (1998) developed an approach for internalizing the externalities into private cost functions in order to achieve a social optimum (example cited from Pearce and Turner (1990)). The differing social and private production costs of a good are the reason why a producer will use the resource more than he would if he had to bear the additional costs that society is faced with. As a result of the overuse of the resource, the social benefits are lower than the social optimum, as the additional external costs decrease the aggregated welfare.

Figure 4 shows how the external costs can be integrated into the decisions of a private producer. Here, a production intensity that is strongly related to soil erosion is assumed and high outputs mean high rates of erosion measured in tons per ha. The erosion rate does not affect the production costs of the producer, so the optimal level of production for the producer is where his marginal net private benefits (MNPB) equals zero, given a certain price level. This is the point where the increase of one unit in production will not increase the profits of the producer. The producer realises an output based only on his private optimum (MNPB=0).

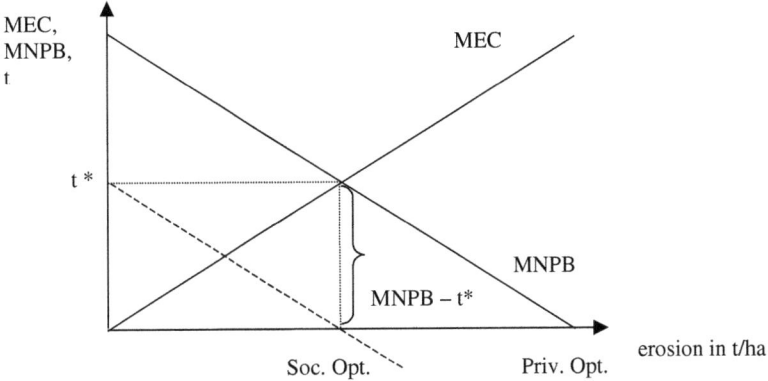

Source: Pearce and Turner 1990

Figure 4: Optimal pollution tax t*, marginal external costs (MEC) and marginal net private benefits (MNPB)

Society is affected by the increased soil erosion rate, which leads to increased marginal external costs (MEC) per unit produced.

However, if the social optimum is reached, the marginal external costs would equal the marginal net private benefits (MNPB = MEC). If it is possible to decrease the soil erosion to this level (Soc. Opt.), the gains through higher benefits to the society would outweigh the losses caused by reduced production for the producer.

Pigou's idea was to introduce a tax t on the product (in this example, the agricultural output) through a social planner (the government) so that MNPB are reduced by the amount of t*, which is exactly the amount of MEC at the social optimum. The producer now re-adjusts his production level given the tax t* and will produce at the socially optimal level (MNPB – t* = 0).

The damage caused by erosion gets reduced to the point where the positive effects of less erosion are outweighed by losses on the producer side.

The example can be transferred to a subsidy that is paid for a production practice that causes less erosion than the standard practice, but which involves higher costs to the farmer. The less erosive practice becomes more advantageous and is therefore more applied than without the subsidy.

4.4.2.2 Negotiation solutions

Coase (1960) questioned the practicability of Pigou's approach, since it might not be possible to define the optimal amount of tax through a governmental agency. In his argument he left it to the individuals involved in the decision problem and extended the approach by introducing the existence of property rights. He made clear that the solution of such a problem is highly dependent on the a-priori-distribution of property rights. The person who has the right to pollute can ask for compensation for the reduction of the damaging activity or if the affected person has the right of not being damaged by the producer, he can sell this right to the producer for the negotiated price. No matter how the rights are distributed at the beginning, negotiations will lead to a pareto-optimal solution. The same example with one producer (a farmer) and a person suffering from the effects of soil erosion is assumed here (representing the society as a whole) (see Figure 5).

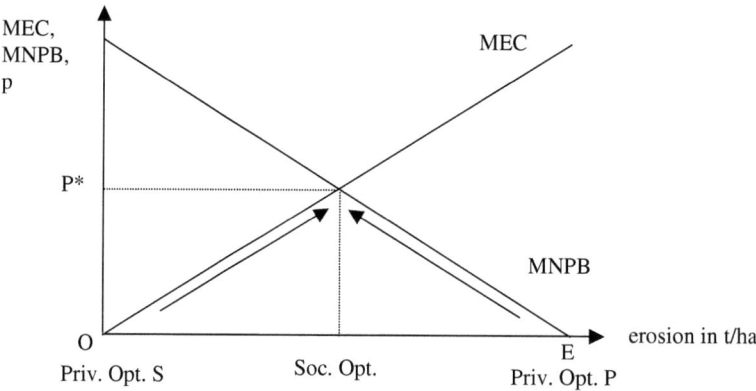

Source: based on Coase 1960, following the example in Pearce and Turner 1990

Figure 5: Social optimum achieved through negotiations

Depending on the distribution of the property rights, there are two possible points where the negotiation can start. If the farmer has the right to pollute, he will produce at the point E (Priv. Opt. P). For the society there is an incentive to pay the farmer for the reduction in soil erosion because it will decrease the marginal external costs more than the decrease of the net private benefits. Given complete information about the costs and benefits of both sides, a price of p* per unit of decreased pollution (ton of eroded soil per ha) will develop, where marginal external costs will meet marginal net private benefits.

If the rights of not being polluted belong to society, zero pollution would take place at the starting point O of the negotiation (Priv. Opt. S). Marginal external costs are zero but the farmer has an incentive to purchase these rights from society (known as the "polluter-pays-principle"), since any increase in production will increase his profit. Here, the bargaining process will end at the same price p*, the only difference is that the compensation is given to society.

Coase made clear that the distribution of property rights is the crucial point in this discussion. Property rights generate income to the person who owns it but negotiations will finally lead to pareto-optimal solutions. The distribution of the property rights only determines which person will start the negotiation process. External effects could be internalized through negotiations between the market participants, so that the (inefficient) implementation of a Pigouvian tax by a government is not needed.

This theoretical approach is not often directly used as a policy instrument. However, it helps one understand how environmental policies are implemented under different institutional environments. If the property rights are mostly claimed by the user of a resource, then the individual or society that suffers damage from the resource use will have to buy out some of these property rights (i.e. incentives or compensations). On the contrary, if the society holds the property rights, then the potential user of the resource will have to pay for "pollution rights" in taxes.

However, Coase made two crucial assumptions for the functioning of negotiations:

1. The transaction costs of negotiations have to be low enough for the process to start.
2. There is complete information and transparency regarding the costs and benefits involved.

Both assumptions are difficult to fulfill, even for a two-person example. Transaction costs are never at zero in real world conditions (Challen 2000). As a result, negotiations would not be possible because of the prohibitively high costs for bringing the stakeholders together. Private information however can be used strategically to improve each person's result (Arrow 2001). Given the problem of soil erosion with a huge number of pollutants and individuals affected, the theory is overrun by real world conditions. The common problem of non point source pollution (Griffin and Bromley 1982) will increase the probability of free riders that will take the compensation but not fulfill their part of the contract.

4.4.2.3 Remarks on the approaches by Pigou and Coase

Both Pigou and Coase developed ways to internalize externalities. Environmental taxes and subsidies (a negative tax) are used in policies for the management of natural resources, e.g. taxes on fossil fuels or subsidies for environmental friendly production (e.g. reduced tillage programmes). In other cases, it is left to the individuals to negotiate for the optimal price of pollution rights. One possible application is seen in the auctioning systems where the price is generated by supply and demand (e.g. the U.S. acid rain program (EPA - US Environmental Protection Agency 2006)). Note that even this example is weak, since the auction system needed governmental input for creating such markets. The US Conservation Reserve Programme (CRP) is an auction example for soil conservation that uses a bidding system to find appropriate land for this programme (Plankl 1999).

The problem with both approaches is the empirical difficulties in gathering knowledge and data for all the involved costs and benefit functions (Pearce 1993). What seems clear in theory requires a lot of effort in real life. Companies may not easily provide confidential information regarding their production process, individuals may exaggerate their suffered damages when compensations are available (asymmetric information) (Arrow 2001). Even though the external costs of pollution exist, it is difficult to relate damages to specific activities (non point source pollution) (Segerson 1988). Weersink et al. (2002) pointed out the impossibility of gathering this information for practical studies.

The above example was only abstracted to a two player case – one polluter "versus" society. If this simple case would be extended to a larger number of individual polluters, cost functions among them would differ causing greater difficulties in generating an aggregated cost function. Further aspects had been derived in Pearce and Turner (1990).

The criticism regarding the assumptions of welfare economics will not be extended here (Kula 1992; Pearce 1993; Pearce and Turner 1990; Pearce and Warford 1993; Tisdell 1991). Nevertheless, the underlying assumptions of the *homo oeconomicus* are difficult to apply to real world problems. However, taxes, subsidies and, to a certain part, negotiations play a considerable role in the real economic world. Even if the functions are not known and the information on the private costs of pollution abatement is sparse, decision makers still use such instruments (not only in a trial and error way) to change and influence the behaviour of individuals. Both Pigou and Coase provided a theoretical concept that is useful for the development of practical implementation options for soil conservation.

4.4.3 Spatial targeting and regulation areas

When environmental programmes are seen as a possible solution for natural resource management, the spatial targeting of such programmes is often suggested (*perfect mapping*) (Breton 1970). The term "*regulation area*" is relevant in the context of agro-environmental policy making (Latacz-Lohmann 2001), while the term *perfect mapping* was introduced by Breton in connection with the general provision of public goods (Breton 1970). The effect of such spatial targeting of environmental programmes had been widely discussed in economic literature (see examples Lankoski and Ollikainen 2003; Latacz-Lohmann 2001; Yang et al. 2005b). In most cases, given the heterogeneous distribution of the environmental assets, targeting was preferable to non targeted policies. The approaches that were used for the selection of areas for soil conservation programmes were usually based on the statistical analysis of GIS data. This chapter provides some information on the spatial effects of the selection of the regulation areas of soil conservation policies. Several approaches are available to help in the selection of land for targeted conservation programmes. However, it is finding the "optimal" choice that poses a challenge. Here, the economic background of spatial aspects for policy instruments (e.g. targeted conservation programmes, regulation areas) are discussed.

The importance of spatial aspects in non-point source pollution was emphasised by Wossink et al. (2001): Heterogeneity in the economic and ecological attributes must to be reflected in the underlying data. Aggregation of spatial data will lead to the false estimation of pollution prevention costs. Lintner and Weersink (1999) showed the effect of the spatial location of farms in a watershed with respect to achieving optimal abatement of pollution, which is an aspect often neglected in the economic analysis of environmental problems.

Being different from neoclassical economics, where spatial effects are more or less neglected, the efficiency of agri-environmental policies is not only dependent on the choice of the appropriate instrument. Whether a policy is suitable and efficient depends greatly on the regulation area: the spatial unit where the policy will finally be legally effective. The choice of the regulation area is important because different agri-environmental problems have different spatial dimensions, ranging from small-scale local problems to issues of national or even global importance. In order to find an efficient solution for such problems, the geographical delimitation of an agro-environmental programme should fit the spatial dimension of the problem in question (Latacz-Lohmann 2001; Scheele et al. 1992).

There is a trade off between costs arising from setting a programme on too large an area (i.e. profit loss in areas that are not actually part of the environmental problem (Yang et al. 2005b)) and costs that arise from a highly detailed scheme that require high administrative efforts for the implementation and control of such a policy (Urfei and Budde 2002). A balance should be reached for these costs through finding the optimal size for the regulation area. This can be achieved through a real world experiment, where the targeted area is increased step by step and the costs and resulting effects are compared (trial and error). The other option is to calculate the optimal regulation area size using a modelling approach, where the objective function maximises the overall net benefits of the regulation area size or selects areas with the highest net benefit.

Figure 6 shows a soil erosion risk map of the two districts Uckermark and Barnim in the state Brandenburg as an example of varying administrative and ecological purposes. Erosion risk zones are scattered within municipalities and cut across borders. A policy that covers either a whole municipality or even both municipalities would create efficiency losses through regulating areas with low erosion risk.

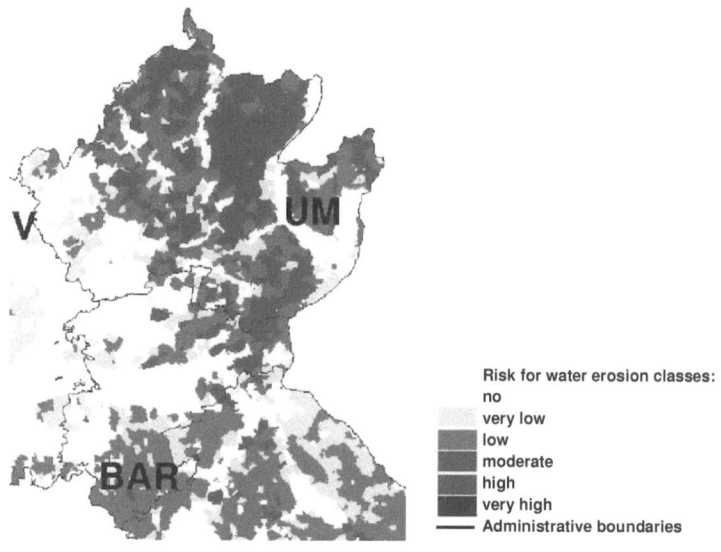

Source: Matzdorf et al. 2003

Figure 6: Administrative and erosion risk units over two districts (UM Uckermark; BAR Barnim) in Brandenburg

Administration levels for regulations, negotiations and agreements can also vary from international level to the very small scale level of a land parcel that indicates the ownership of a certain piece of land. Each environmental problem may need a different regulation level. Some global environmental problems demand solutions on the international level (e.g. greenhouse gas emissions), while the protection of environmental sensitive areas may demand regulations that target only a specific field.

Additionally, ecological boundaries are not as sharp as those set in administration. For example, the boundaries of landscapes can only be vaguely identified or the diffusion of polluted air spreads randomly over a certain area.

For further reading on the aspects of regulation areas, see Rudloff and Urfei (2000); Urfei (1999); Urfei and Budde (2002); Latacz-Lohmann (2001); Scheele et al. (1992).

An approach that is widely used (mostly in the current EU-conservation schemes) all over Europe focuses on entire administrative regions or federal states. This choice is usually made for reasons of equality and easier administration. The regulation area is then tied to administrative boundaries. Whether this choice is suitable in terms of the programme's cost-effectiveness or in terms of ecological results is discussed as one of the subjects of this study, see also (Huylenbroeck and Whitby 1999).

Agricultural administrations in the EU are up to now not very experienced with targeted conservation programmes, for most existing agri-environmental programmes are based on a standardised action oriented approach (Huylenbroeck and Whitby 1999). Exceptions are seen in only a few very site-specific contractual programmes for nature preservation that are undertaken with the cooperation of environmental agencies (dt. *Vertragsnaturschutz*) and some result-oriented approaches such as the MEKA programme in the German federal state Baden-Württemberg or in agri-environmental programmes in Switzerland (Oppermann and Gujer 2003).

Within this study, the influence of different selection criteria on the size of eligible area for targeted conservation programmes will be demonstrated. Then, within a chosen regulation area, different options of conservation programmes will be analysed. This is done through a combination of measures and instruments. The question is whether it is preferable to use a policy with its instruments on a whole region while accepting some efficiency losses or to focus the policy only on highly erosive areas, which might be more efficient but accompanied by higher transaction costs.

4.5 Methodological approaches for the economic analysis of soil conservation policies

The following chapters give an overview of approaches for the economic analysis and ex ante evaluation of policy options, with the aim of preventing policy failure. The basics of cost-benefit-analysis including the concept of Total Economic Value (see Table 5) are discussed as a tool for the economic analysis of conservation policies. Then, cost-effectiveness-analysis is introduced as a solution for decision problems, where the costs and benefits of the environmental resource use are difficult to measure. The cost-effectiveness-analysis serves as a more pragmatic approach, which is combined with the concept of "safe minimum standard of conservation" (Ciriacy-Wantrup 1963).

4.5.1 Cost-Benefit-Analysis

Cost-benefit-analysis is a basic analytical instrument based on welfare economics for the evaluation of projects. It allows for the comparison of net effects of policies by using the theoretical background of welfare economics and by accounting the costs and benefits of a policy option in a static way.

Cost-benefit-analysis (CBA) is commonly used to evaluate public development projects. It searches for the implementation option that brings the highest net benefit for society, which is an often needed governmental decision support for public spending (Musgrave and Musgrave 1989). Hanusch (1987) described cost-benefit analysis as a tool for the "better" provision of public goods. The term "better", as a normative expression is defined as the more efficient use of public money, expressed as the above mentioned net benefit of a project or programme (Hanusch 1987).

Cost-benefit-analysis should examine all costs and benefits that are accrued to and result from a project that can affect on a national or even international level, including those costs that cannot be measured or valued (Clark 1996). An attempt was made using CBA to assess the economic impacts of projects or policies by analysing and comparing the total costs and benefits instead of finding the cost and benefit functions related to environmentally hazardous activities (based on the production function). This has the advantage of being able to compare the two scenarios of a development (i.e. with or without a certain policy or project). However, cost-benefit-analysis as a partial analysis instrument, is not capable of reflecting dynamic changes in the preferences over time (Tisdell 1991).

CBA may not necessarily find the optimum level of pollution but can give information on the net benefits of different policy options (Hanusch 1987). This process can be described by comparing two points on the benefit and cost functions, where the shape of the curve is not known but certain points can be described. It is more adapted to reality, where possible solutions to a problem are first developed within a political process. Only these prototypes are then assessed using CBA.

As for soil conservation, Crosson (1984) posed the fundamental question of whether tolerable soil loss rates should be used and whether future generations should face higher costs. He pointed out that a loss in productivity can be levelled out by better technology at lower costs than investing everything in erosion control now. According to Crosson, "the key is the relative cost of these two alternatives", which depends highly on the measures used and the economic damage caused by erosion. Therefore, one option is to not avoid erosion but to avoid the impact on valuable water bodies through measures like riparian buffer strips.

The valuation of the costs and benefits has to be based on the opportunity costs of the produced and used goods and services, which make up three distinct groups (Clark 1996):

- Traded goods for which a market price exists can be valued at their world market price (adjusted for the transport and distribution costs)
- Domestic factors of production (i.e. land and labour) are valued with their marginal value product (e.g. market prices, market rents)
- Non tradable goods and services should be split up into their tradable and non tradable components. The tradable parts can be valued at market prices, whereas the others are valued at the marginal value product or the marginal social benefit (e.g. willingness to pay).

For some costs and benefits any evaluation is almost impossible. In these cases a qualitative listing and description is preferred over completely excluding these factors.

The setting of the discounting rate and the time frame of assessment that is used to adjust future cost and benefits on a common basis is an important aspect in cost-benefit-analysis (Pearce and Turner 1990). It is crucial to explain the viewpoint of the analysis, whether it is from an individual or from society's perspective. The discount rate is oriented either on a social discount rate (i.e. the opportunity costs of capital in the public sector) (e.g. (Abelson 1979)) or higher discounting rates when the project is evaluated more from the viewpoint of a farmer that has to assume interest rates based on bank loans (Clark 1996) or flexible discount rates over certain time spans (Pearce et al. 2003) (see also Chapter 4.2.3.2).

In environmental economics, valuation concepts are still being developed and improved so that the concept of standard cost-benefit-analysis for public projects may be extended (Turner et al. 2003). The evaluation of public goods produces theoretical difficulties because these goods are not usually traded on markets for the above mentioned reasons (e.g. non-excludability, non-rivalry of consumption) and even if they were, the prices would be distorted for the same reasons (Tisdell 1991). Therefore, more information needs to be gathered to define the value of a natural resource, so that the economic value of a natural resource may be found.

Since the over-emphasis on measurable economic values of development projects was such a main criticized point in standard cost-benefit-analysis (CBA), and the more ecological, "soft" values were ignored, the **Total Economic Value** (see Chapter 4.1 p. 29) as a more comprising approach was introduced (Pearce and Turner 1990; Turner et al. 2003). In environmental economics, the range of values that must be taken into account had been more and more extended while more methods for measuring non-marketable values had also been developed.

The actual estimation of these values is achieved with the help of valuation techniques such as the willingness-to-pay (WTP) or the willingness-to-accept (WTA), dependent on the distribution of property rights of the natural resource. Both methods are based on surveys of involved individuals to help find out the amount they are willing to pay to preserve or conserve a certain condition or to access a natural resource or if the property rights are distributed reciprocally, what amount they would demand as compensation for the loss in their property rights (quality of their property).

Empirical methods used in this field are the travel-cost method, hedonic prices, conjoint analysis and contingent valuation (see Pearce and Turner 1990).

An example of the application of contingent valuation to soil conservation is given in Colombo et al. (2003). In this study, the benefits of a soil erosion control program for the general public were estimated. The willingness to pay for erosion reduction in the specified area was estimated at around 42-72 €/hectare/year. Furthermore, the authors brought up the significant variability in the value placed on soil erosion control when the respondents were reminded about substitute environmental projects.

This behaviour is one of the usually criticized points of most WTP approaches. Most studies focus on a single project, so the results are not related to the other expenditures of the individuals involved in the survey (Turner et al. 2003). Reminding the respondents of other options that also could use financial support will decrease the amount spent on a certain good (*embeddedness*) (Cummings et al. 1994; Edwards and Anderson.G.D. 1987).

The described approaches and methods are focused on estimating the value of a certain resource. On some occasions, it can be assumed that society will assign a certain value to soil conservation in the form of preferences towards a specific policy.

The valuation techniques demand a lot of effort for collecting information on the willingness to pay or to accept but the results are still faced with criticism due to the shortcomings and variability of the used methods (see above, (Cicchetti and Wilde 1992; Lazo et al. 1992)).

Although CBA for soil conservation programmes has the potential of providing useful information on the effects of a policy, the difficulties in estimating the total economic value of soil and soil erosion can create bias in the outcome of a CBA. Definition problems surrounding the range of affected individuals may add to the ambiguity.

4.5.2 Cost-Effectiveness-Analysis based on a Safe minimum Standard

The cost-effectiveness-analysis (CEA) is the appropriate tool for comparing different options for achieving a certain specified objective without the difficulties faced in cost-benefit analysis (interest rates, methodological problems finding the WTP) (Levin and McEwan 2001). It is not possible to analyse, whether a project has a net benefit to society with CEA, but given a certain agreement within society that soil conservation needs public intervention, the most cost efficient solution can be found.

There are approaches that theoretically lead to an optimal solution but lack in feasibility under real life conditions. However, an approach based on the works of Ciriacy-Wantrup (1963) must be emphasised, as it allows a theory-based application within cost-effectiveness-analysis. Ciriacy-Wantrup (1963) developed this economic conservation theory, which he described as the "safe minimum standard" (SMS). The following chapters will outline the CEA and the concept of safe minimum standard.

4.5.2.1 Characteristics of CEA

For the CEA, not only are the net benefits (benefits minus costs) of two or more options compared, but also the costs that arise from achieving a certain goal. Therefore, the result of the CEA is the ratio of goal achievement in relation to the costs or vice versa. In the example of soil conservation, the cost-effectiveness shows how many tons of erosion are avoided per unit of money spent in conservation programmes or the costs of avoiding one ton erosion through a specific programme.

Cost effectiveness ratio = $\dfrac{\text{tons of erosion avoided}}{\text{costs of programme per hectare}}$ = tons/€/ha

Or: $\dfrac{\text{costs of programme per hectare}}{\text{tons of erosion avoided}}$ = €/tons/ha

The costs of a programme can be derived from the loss in the total gross-margin of a given area and/or the budget costs that arise from incentive payments. This study analyses the opportunity costs of farmers for providing or adopting soil conservation practices. In the case of incentive payments through a conservation programme, these costs have to be added to the gross margin changes of the region. The costs of erosion abatement can also be shown by the shadow price of one erosion unit in an optimization model, if the erosion variable is limited in a model through a restriction.

4.5.2.2 Definitions

There is a multitude of definitions for the terms effectiveness and efficiency; the differences in meaning within the English and German vocabulary only adds to the confusion.

In order to avoid any confusion in the use of these terms, some working definitions for these terms are given below:

Efficiency describes the economy of an action. The German *Fischer economic dictionary* (Rürup et al. 2002) defines efficiency as the "economic utility of a given situation; with efficient decisions the economic principle is implemented." In the context of soil conservation, the economic efficiency describes how many tons of eroded soil can be avoided with a given budget or as mentioned above, a given protection level is implemented at a certain cost.

Effectiveness describes the ability to achieve stated goals or objectives, judged in terms of both output and impact (Bureau of Justice Assistance 2006). The ecological effectiveness of soil conservation measures quantifies the ability or the degree of contribution of a measure to soil erosion prevention (e.g. in tons/ha/year).

To clarify the terms for this study the term *efficiency* is reserved for the economic view, while *effectiveness* describes more the physical terms. *Cost-effectiveness* refers back to efficiency, as it describes the effectiveness of an action in monetary terms. It is the effectiveness of a certain measure related to its costs (Clark 1996; Levin and McEwan 2001). However, since cost-effectiveness-analysis is a commonly used term in economics, it will be used as it is. Efficiency is described as a cost-effectiveness ratio.

4.5.2.3 The Theoretical Background of Safe Minimum Standard

In Ciriacy-Wantrup's book on the safe minimum standard of conservation (SMS) (Ciriacy-Wantrup 1963), he underlined that uncertainties exist for the physical characteristics and behaviour of natural resources. Given this uncertainty, any optimization procedure would be only valid for one spot in a huge field of uncertain conditions. Therefore, it is justified in natural sciences to find a threshold or to formulate a goal for conservation, a safety zone between the one spot optimal solution and the uncertainty of other outcomes. Ciriacy-Wantrup stated in order to maximize welfare such resources cannot be managed based on an optimal management approach or a pure cost-benefit-analysis that reflect the discounted present value.

Ciriacy-Wantrup (1963) defined the SMS as a physical term: "... as a flow rate, as specified physical conditions necessary for maintaining such a rate through unspecified conservation practices, or in terms of performance of specified practices. In this sense, the safe minimum standard may be regarded as a technological constraint in economic optimizing."

However, Ciriacy-Wantrup specified later that the SMS approach can also be defined as part of the "objective function" rather than part of the technical constraints. "In this respect the SMS is more akin to an institutional than a technological constraint" (Ciriacy-Wantrup 1963).

Toman (1994; cited in Pezzey and Toman 2002) underlined the appropriateness of an extended SMS as a link between efficiency considerations in cost-benefit analysis for investments and ethical values that come into play with long term, intergenerational equity: "Standard trade-off [cost-benefit] analyses apply when the magnitude and duration of risks are not very large, so moral stakes also are relatively low; however, ethical norms become increasingly important complements to trade-off analyses as the stakes rise". This is especially true for soils: the correlation of soil use and long term fertility of soils is affected with uncertainty, soil replenishment rate are not known. Therefore, a safety zone above an assumed "optimal" soil erosion rate is recommended.

In other words, SMS might not be a "straight forward" economic tool, but it represents a behaviour that is very common in most decision making processes, namely the setting of binding limits, regulations and thresholds through a political process with the help of policies and laws.

In Ciriacy-Wantrup's (1963) argumentation, he wrote it is not a philosophical reason that calls for such a standard but the mere uncertainty that accompanies physical, biological and ecological processes. As has been described before, only risks can be approximately measured or defined, whereas uncertainty is beyond such measurability. Therefore, an economic valuation influenced by uncertainty would not deliver any useful results.

Other authors had added further justifications for the use of SMS. Most of these authors developed the philosophical position that says there are limits beyond which utilitarian calculus ceases to be legitimate. Bishop (1978; 1979) assumed that planners are unaware of the probabilities of relevant events. Therefore a planner should follow a "minimax" strategy, which minimizes the maximum possible loss. "This strategy is tantamount to assuming the worst possible outcome is a certainty, and will not in general maximize welfare" (Margolis and Nævdal 2004).

Since it is mostly questionable as to where to draw the line for a safe minimum standard, Bishop (1978) introduced a decision rule, where the option for conservation should be maintained until the costs of conservation are prohibitively high. An extreme example would be that the extinction of any species is not acceptable, unless the survival of entire mankind is threatened (see also Bishop 1979; Crowards 1998).

4.5.2.4 Implementation examples of a safe minimum standard

Most empirical examples used with SMS deal with the extinction of species, e.g. a fish that became extinct after the building of a dam, or the last Californian Condors endangered by oil mining. However, Ciriacy-Wantrup (1963) explicitly gave the example of soil use regulations that were in fact functioning as a safe minimum standard. The implementation of a SMS is done by introducing either a threshold or a goal as an exogenous restriction in the decision problem, which in fact is applying (and socially accepting) the above mentioned technical constraint on the use of a resource.

4.6 The application of CEA with a Safe Minimum Standard

For this study, the Ciriacy-Wantrup statement of a safe minimum standard as a threshold or a safety zone between the one spot optimal solution and the uncertainty of other outcomes is rather adequate. As stated before, the replenishment and erosion rates of soils are characterized by uncertainty. Therefore, limiting soil use to a socially agreed-on standard is more advisable than finding an economic optimum that might bear the risk of complete destruction of the resource (Dabbert 1994). The use of a safe minimum standard that is not based on economic calculus but that allows the sustainable use of soils for more than one generation offers intergenerational equity without the question of finding the "optimal" interest rate for the discounting of future profits. Real world problems are never only a pure economic problem, since they usually involve legal and social spheres as well. These spheres are linked by institutions such as property rights, which govern the use of any resource based on traditions, laws and moral boundaries (Tisdell 1991) (see also Chapter 8).

In the case of this study this approach translates into the limitation of soil run-off proposed by soil science, but which is assumed to be negotiated and implemented through a political system. The role of economics is then to analyse the costs that come with achieving such goals and find options that minimise these costs (cost effectiveness analysis). In other words: The approach used in this study starts at a point, where the need for conservation is revealed by society through the time spent on the issue in a political process, the political will to formulate relevant regulations and public money provided for the management and support of soil conservation programmes. The task is then to find efficient implementations of conservation programmes for a given budget.

The safe minimum standard can be used to define the goal to reduce soil erosion within an area to a specific amount (see chapter 7, p. 98). The SMS is then implicitly formulated by a maximum total sum of eroded soil that should not be surpassed within a region. The safe minimum standard can also be used as a selection criteria for eligible area (*spatial targeting*) for soil conservation programmes (see chapter 5.4.1, p. 71) when thresholds for potential soil loss are applied.

4.7 The analysed implementation options

4.7.1 Definition of instruments and measures

Within this study, *measures* and *instruments* are distinguished following these definitions. A *soil conservation measure* defines the practical action or the production level of the farm (Bridges et al. 2001), while an *instrument* is the tool on the policy level that influences the farmer's actions through financial incentives, legal regulations or farm extension to implement (or to avoid) certain *measures* (Weersink et al. 1998).

The analysis of soil conservation measures must enable both predicates in terms of the ecological effectiveness and the economic efficiency. The effects of measures and instruments cannot be regarded in isolation from each other but instead must be combined into a consistent evaluation framework. Therefore, the economic analysis of both the measures and the assigned instruments is necessary.

In the following, implementation instruments (the policies) and the on-farm measures will be described. Finally, among these alternatives an initial selection will be made for further analysis in the empirical part of this study.

4.7.2 Instruments

In Chapter 3.2 and 3.3 an overview of the available instruments for promoting environmental goals was given for Germany. Chapter 4.4.1 described these instruments from an economic viewpoint.

Environmental economic *instruments*, which are applicable to the agrarian sector comprise legal regulations and prohibitions, incentives or taxes as well as certificates (Weersink et al. 1998).

Of the given alternatives, two general options will be analysed in this study:

1. an **incentive option**, using subsidies for certain conservation measures and
2. a **legal regulation option** based on preventive prohibitions with allowed and prohibited measures on highly erosive land.

The other options are not analysed for the reasons below:

Planning instruments are not suitable since they are used more for general decisions concerning land use (i.e. zoning for residential or industrial areas). This instrument only applies on a much higher level from a hierarchical view on soil conservation approaches. Changing the shape of fields involves a much higher number of stakeholders than only the land users and the governmental agencies (Arzt et al. 2003). Especially in Eastern Germany, each field unit has more than one owner, even though the field is cultivated by only one farmer. Changes in the field shape or the set-up of hedgerows require the permission from several owners, which is unpopular and uncommon in this region (Arzt et al. 2002). Additionally, this option needs a different research approach which goes beyond the scope of this study. Regulations can have similar effects as planning instruments when they are used in a site-specific way. A regulation that prohibits soil damaging activities on highly erosive sites is indeed a spatial planning tool. For this study, these aspects are analysed in the example of a regulation instrument. The application of an environmental impact assessment is not suitable for soil conservation policies, since it is only used for the evaluation of environmental effects of large scale construction projects such as highways or electric power plants (Schmidt and Müller 1992).

Thus, the two selected options represent appropriate ways in which soil conservation can be promoted under the conditions of a given set of property rights and within a short time horizon (Tisdell 1991). The incentive option functions in an indirect way as a negative tax in the *Pigou* sense. This can be seen as accepting the existing set of property rights combined with temporally selling the right of use, namely the right to use the resource in a formerly agreed-on way (Tisdell 1991).

The environmental effect of a legal regulation might be clear and predictable. Depending on the authority's capabilities in control, the compliance to such an instrument might be questionable (Latacz-Lohmann 2004). From a property rights point of view, a legal regulation is a partial transfer of property rights, i.e. the right of use is changed to less damaging level.

Voluntary incentive options have less foreseeable effects, since the adoption of such instruments is dependent on many factors that go beyond the level of the given subsidy (Drake et al. 1999). However, the resistance against the implementation of such instruments is usually lower than those against legal regulations: if the instrument is not suitable, the farmers will not adopt it (Falconer 2000; McCann and Easter 1999a). Economic factors related to adoption can be modelled based on average values (Baudoux 2001; Schuler and Kächele 2003). Factors based on attitudes are difficult to model, they require empirical surveys (Drake et al. 1999; Falconer 2000; McCann and Easter 1999a). Such resistance will be addressed under the chapter on transaction costs (see Chapter 8).

4.7.3 Measures

The practical *measures* examined in this study that serve as variables for the prevention of soil erosion comprise:

- the cultivation of crops with a lower erosion risk,
- cultivation of cover crops,
- under-sown crops,
- reduced tillage systems.

Crops themselves vary highly in terms of their erosivity. Highly erosive sites can be converted from arable land to grassland, which has the lowest soil erosion risk among all the production activities. Corn has higher erosion rates than winter wheat. Other options for reducing run-off from agricultural fields are the use of intermediate and under-sown crops. Reduced tillage systems leave the soil more covered with residues, thus reducing the risk of soil getting washed away by rainfall. The actual values for these production alternatives are shown later in a separate chapter (see Chapter 5).

The re-designing of agricultural areas can also help prevent erosion enforcing field shapes (i.e. with long slopes). However, the change of field-size, shape or the building of terraces is not implemented in the model because such changes usually demand higher efforts in planning (see instruments) and are not comparable to on-site measures e.g. reduced tillage. The feasibility of such measures are usually challenged with strong resistance in the given property rights framework (Arzt et al. 2003).

The final model set up will be described in more detail in Chapter 6. At this point, a theoretical approach will be developed, which will lead to a cost-effectiveness-analysis based on a safe minimum standard.

5 Soil erosion risk assessment of sites and cropping practices – the effectiveness of soil conservation measures

For the cost-effectiveness-analysis of soil conservation measures, two parameters are used, which describe the soil erosion risk of each agricultural measure on different sites so that the performance of a measure is related to its costs. The two parameters for the erosion assessment are:

1. a parameter describing the potential erosion risk of the agricultural land (erodibility) (Chapter 5.1),
2. a parameter that describes the erosivity of each agricultural practice (Chapter 5.2),
3. and a combination of both parameters showing the potential erosion risk of a specific practice on a specific site (Chapter 5.3).

As a precondition, a model that provides this soil erosion risk assessment based on data that are easily available from GIS sources (which describe soil qualities, land use patterns and elevation maps) is needed.

The soil erosion model itself provides in a first step separate values for the site erodibility (e.g. the natural conditions of the area and the corresponding risk) so that areas with an elevated erosion risk can be found (1). In the second step, the effects of management decisions such as the crops chosen and the tillage practices are evaluated with a more specific assessment model based on a fuzzy logic tool (see Sattler (2007)). This parameter describes the erosivity of each agricultural practice on a specific site serving as a parameter in the bio-economic model (2). The combination of both parameters is shown in chapter 5.4. As an example, the effects of thresholds that were applied on the site specific soil erosion risk data for the selection of eligible areas in soil conservation programmes is described in chapter 5.4.

5.1 The potential erosion risk of the agricultural land

5.1.1 Erosion models for estimating soil erosion risk

This study's focus is on soil degradation caused by erosion through run-off. Therefore, information on the specific soil erosion risk is needed. Several models had been developed and applied to fulfil this task. Bork (1991) had distinguished these models into three groups according to the way soil erosion processes are described:

1. Empirical soil erosion models
2. Deterministic-analytical models
3. Dynamic-deterministic-numerical models

More details will be shown in the following chapters.

5.1.1.1 Empirical soil erosion models

This group comprises pure empirical models that usually consist of simple regressions that describe soil erosion as a result of empirically measured soil losses, calculated with a limited number of parameters. These models are based on the correlation of measured erosion values and easily measured information i.e. precipitation, soil, slope, vegetation and tillage. Input data needs are rather low. Such models are generally only applicable in areas they are calibrated for, so the transfer of these models into other regions is not appropriate. Even though these models are rather simple and limited in their temporal detailedness of soil loss events (i.e. usually average values for one year), they are widely used to help choose eligible areas for soil conservation measures. Typical examples are the "Universal soil loss equation" (USLE) (Wischmeier and Smith 1978) and the German adaptation "Allgemeine Bodenabtragsgleichung" (ABAG) (Schwertmann et al. 1987). Even though these models were developed for use at the field scale level, they have been successfully applied for the soil erosion risk assessment of larger areas (Renard et al. 1991). The original USLE was later improved in some of its factors and named as the revised Universal Soil loss equation (RUSLE) (Renard et al. 1991).

5.1.1.2 Deterministic-analytical models

A second group of soil erosion models is described as deterministic-analytical models (Bork 1991). These models are based on general rules (e.g. physics, chemistry) that are used to describe the erosion processes with simplified mathematical equations. An example of this model type is the Water Erosion Prediction Project (WEPP) (Laflen et al. 1991). Data needs, as well as the computational efforts for these models are higher. To some extent, the transfer of these models to other regions is possible. However, the high effort for model developing and validation, the huge data needs and the longer computation times outweigh the advantages provided by the models, when only an erosion risk assessment is needed. Furthermore, deterministic-analytical models are also empirical models but the theories and rules they are based on are not as simple as in regression models.

Another model type that is widely used in bio-economic models (e.g. Barbier and Bergeron 1999; Deybe and Flichman 1991; Donaldson et al. 1995) is the Erosion Productivity Impact Calculator Model (EPIC) (Williams et al. 1983); a model that includes a plant growth model in its erosion assessment. This plant growth model allows the implementation of the mutual effects of plant cover and soil erosion. Low soil depths decrease the fertility of soils but a dense plant cover of the soil slows down erosion processes. Nutrient loss through soil that is washed from the fields is also considered. Soil properties are characterized by factors taken from the USLE.

5.1.1.3 Dynamic-deterministic-numerical models

A third group is described as dynamic-deterministic-numerical models (Bork 1991), which uses solving strategies from numerical mathematics. These models can give an almost exact description of the erosion processes but given the high data needs and the extremely high computational efforts, these models are only applicable for small parts of a landscape and are not yet suitable for the assessment of larger areas.

5.1.2 The soil erosion risk model in this study - an adapted USLE approach

For the assessment of the soil erosion risk of the study region an adapted approach based on the USLE, which was adapted to the specific research needs of this study was chosen (see Chapter 5.1.1.1) (German version: ABAG (Schwertmann et al. 1987)). The model was chosen for reasons of practicability and sufficient detailedness in the description of the soil erosion risk.

The factors of the USLE can be divided into two groups (see Figure 7):

1) site specific properties (e.g. soil condition, rainfall) and

2) factors that are dependent on the land users' actions and management decisions.

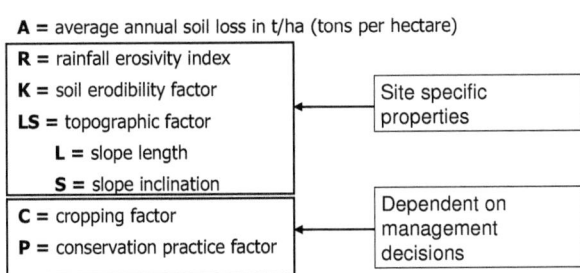

$$A = R * K * LS * \mathbf{C} * \mathbf{P}$$

A = average annual soil loss in t/ha (tons per hectare)

R = rainfall erosivity index
K = soil erodibility factor ← Site specific properties
LS = topographic factor
   L = slope length
   S = slope inclination
C = cropping factor ← Dependent on management decisions
P = conservation practice factor

Source: own presentation; based on Wischmeier and Smith 1978

Figure 7: Universal Soil Loss Equation: Explanation and classification of factors

Of the factors in the second group in Figure 7, cropping factor C is the most crucial factor for this study since it reflects the decision variable within the modelling system, e.g. the crops used and the tillage system applied.

Generally, the USLE is used to calculate site specific erosion values without the distinction of both groups. It is important to note that for this model, both groups are calculated separately in order to distinguish between site specific erosion risk and the erosion risk that is caused by the management decisions. In the following, the site specific erosion risk will first be derived. After this, the influence of the cropping practices will be done. Finally, after the design of the bio-economic model, the combination of both groups will show the potential erosion risk based both on site conditions, crop selection and management effects.

5.1.3 Erosion risk assessment of site conditions

The site specific data was based on research done by Deumlich et al. (1996). Precipitation data is available for the region on a daily basis, soil data was taken from a meso-scale agricultural soil characterization map (Schmidt and Diemann 1981), a map that was used to describe soil quality, slopes, hydrologic properties of all agricultural areas of the former German Democratic Republic. The topographic factor for slope length and inclination was provided by a digital elevation model. Data resolution was based on a grid of 25x25 meter, (see also Sattler 2007). Forests and other non-agricultural area (e.g. settlements) were filtered within the GIS system on the basis of a biotope mapping approach (Landesumweltamt Brandenburg (LUA) 2002).

For the purpose of a general assessment of the region's erosion risk, the C-factor is fixed to the average value (C-factor = 0,11) of the usual crop share that is grown in the region, as a single value that represents the average risk of erosion for the whole region is needed (see (Deumlich et al. 1996)). Note, that this C-factor is replaced in Chapter 5.3 by the values that are derived in a separate process (Chapter 5.2).

The values of the 25 x 25 meter grid cells in the soil erosion assessment map describe the potential soil erosion risk with a regional average rotation under current conditions in the study region (Deumlich et al. 1996) (see Figure 8). This procedure provides a standardised risk assessment of the area without the influence of differing C-factors (potential erosion risk under current agricultural practices).

White areas are forests and settlements that have been excluded from the calculations. Note the heterogeneity of the landscape reflected in the scattered pattern of highly erodible land next to land with low erosion risk values.

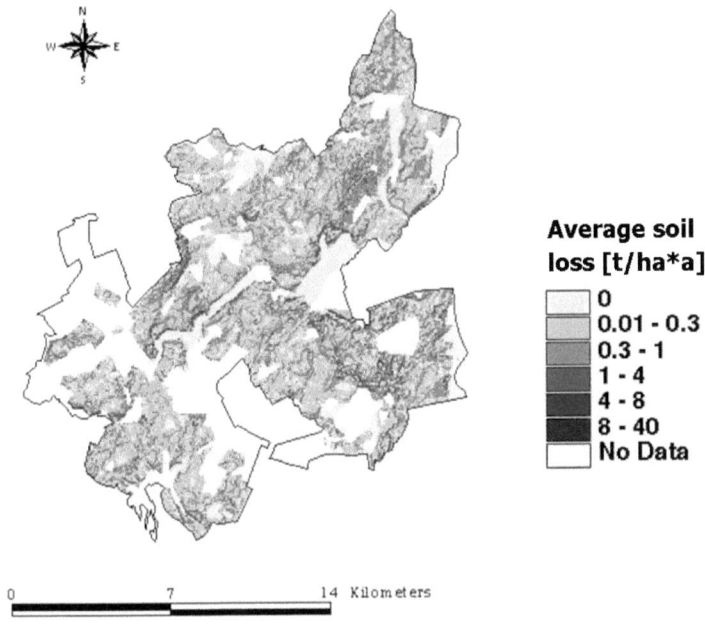

Source: Deumlich et al. 1996

Figure 8: Erosion risk map of the study region grouped according to soil erosion risk categories from Table 6

5.1.3.1 Selection of appropriate soil unit sizes

In order to represent agricultural management units better and to reduce the computational efforts for a regional model, the site-specific erosion risk values based on the 25 x 25 meters cells within the GIS-database had to be aggregated.

Therefore, the original grids were aggregated using three different scale levels. The first scale level was aggregated to grids of 100 x 100 meter grids, representing 1 ha of agricultural area, while a second scale used was formed by grids of 500 x 500 meters. The resulting size of 25 ha is the typical field size in North-Eastern Germany (Werner 2006). Another aggregation method, which is based on GIS-shape files describing the spatial situation of aggregated EU administration units was additionally used (IACS; dt. *INVEKOS*[9]). These units summarise actual land units (dt. *Flur*), which are stored in a database for administering EU-hectare based payments. The advantage of these units is that they are related to real field units, which would facilitate the possible implementation by agricultural administration agencies. Furthermore, data were available from the same source describing the crops grown on these units for the last two years (Matzdorf et al. 2003), based on

---

[9]Integrated Administration and Control System. Combines the electronic coverage, processing, control and payment of EU subsidy application. Dt. *Integriertes Verwaltungs- und Kontrollsystem*.

aggregated project data). These data were also used to calibrate the economic model and validate its results (see chapter 7.4).

5.1.3.2 Statistical choice criteria for the potential erodibility of aggregated grid cells

For the aggregation of the basic 25 m-grid cells a statistical method is needed to give each aggregated cell a value for its potential soil erosion. GIS-software offers two methods for the aggregation of data: either the maximum or the mean value of a group of grid cells describes the aggregated value.

An example for the aggregation options is demonstrated in Figure 9: While the 'max' command uses the maximum value of the grouped cells to define the resulting grid cell, the 'mean' command gives the arithmetical mean to the aggregated cell.

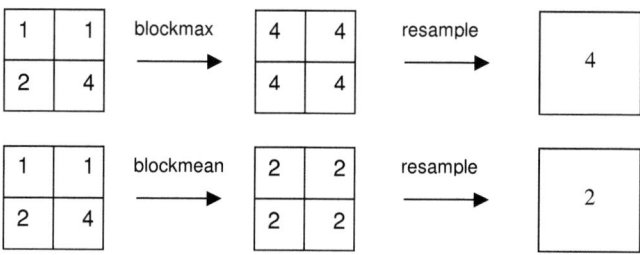

Figure 9: Aggregation commands used in GIS; initial values transformed with 'mean' or 'max', then resampled to new grid cell (Sattler 2007)

Figure 10 explains the reason why different statistical functions had to be taken into account. If a mean value over the sum of erosion risk values would be used, the erosion risk in an area with heterogeneous erosion risk values due to flat and hilly parts within it would be underestimated. The maximum value within a group of grid cells overestimates the risk. However, information that point out areas that show a high erosion potential within an aggregated cell is valuable and should not be allowed to get lost through averaging the risk. A possible solution is to split the aggregation into a labelling and a calculation step. The labelling step shows the method used to generate information regarding the contained aggregated cells (maximum or mean values), while the calculation step produces the expected soil erosion risk as a real value (e.g. tons per hectare). As a result, two types of aggregation were applied:

1.) The statistical mean generates the average value of an aggregated grid cell. This procedure accepts the potential levelling of erosion values between high and low erosion risk cells. Both labelling and calculation for the potential erosion level of an aggregated plot are done with the mean value. The option's acronym is therefore **"meme"** (mean-mean), describing that the mean value is used for both the labelling and the calculation of the erosivity. The pure maximum aggregation method (see Figure 10, both labelling and calculation using the max value) was rejected since it overestimated the erosion risk in the region.

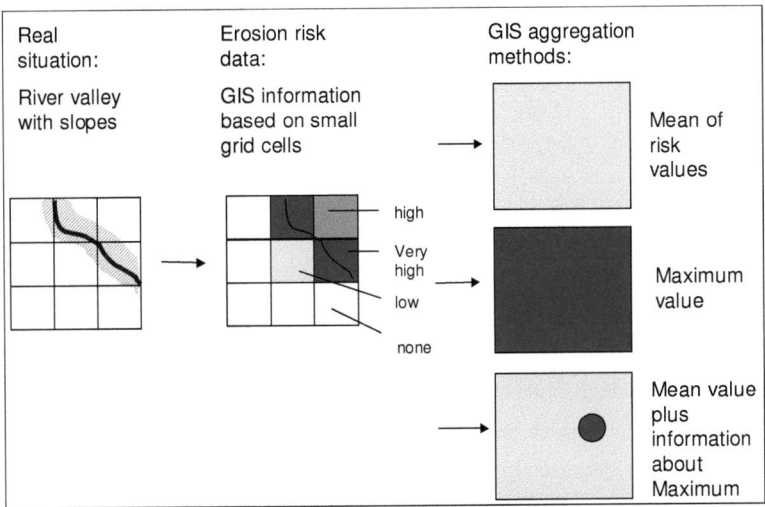

Figure 10: Aggregation methods for soil erosion risk information using different statistical functions

2.) As a synthesis of both maximum and mean method, the second method uses the maximum option to mark cells with a high erosion risk, i.e. the maximum value among the cells that form one aggregated unit. This value is chosen to describe the whole unit as a specific erosion risk group. For the actual calculation of potential soil erosion caused by each cropping activity in the model, the mean value of each cell is used, in order to avoid the overestimation of potential soil depletion. This procedure prevents the levelling around the mean value and covering potential high erosion risk along steep topographic units such as river banks or steep hills, while still showing the actual erosion risk potentials. The option's acronym summarises as "**mame**" (maximum-mean), where "ma-" stands for maximum value used to find high erosive units, and –me is used for calculating potential soil erosion.

5.1.3.3 Clustering into soil categories

The resulting aggregated units were grouped into categories in order to further reduce the number of individual units. The site specific erosion risk values that were calculated individually for every single grid cell of the sample area were clustered in the first step to form 6 erosivity field types (see Table 6) that were based on expert opinion and reflect the heterogenous conditions of the case study area (Sattler 2007). A German study on water protection considered erosion risk of less than 2 t/ha*a as still very low (Frede and Dabbert 1998). However, in this study the lower grouping levels were useful since the averaging effect of the applied aggregation method caused overall lower levels of erosion risk.

Table 6: Categories of the site specific erosion risk for water erosion based on the ABAG-USLE assessment with a standard C-factor (0.11)

Cat.	Risk of water erosion	Soil erosion [t/ha*a]
1	none	0
2	very low	0,01 – 0,3
3	low	0,3 – 1
4	moderate	1 – 4
5	high	4 – 8
6	very high	> 8

Source: Sattler 2007

Then, in the second step, three soil quality classes based on the German soil quality system (dt. *Ackerzahl*) were used to cluster the GIS-soil data of the region. The applied soil quality classes (25, 38, 50) represent the medians of the clustered soil classes. The later described farm model uses such soil fertility classes to distinguish the yield levels. Finally, the soil quality classes are combined with the erosion risk categories to generate 18 soil types (see Table 7).

Table 7: Erosion risk–soil quality types as a combination of soil quality class and erosion risk category

Soil qualiy class (AZKL)	Erosion risk category (acc. to Table 6)					
	1	2	3	4	5	6
25	25_1	25_2	25_3	25_4	25_5	25_6
38	38_1	38_2	38_3	38_4	38_5	38_6
50	50_1	50_2	50_3	50_4	50_5	50_6

Source: Sattler 2007

Table 8 summarises the possible combinations of the two statistical grid aggregation methods and the three land size categories. The first part of the resulting key describes the aggregated grid sizes followed by the key for the GIS aggregation method.

Table 8: Possible combinations of statistical grid aggregation and threshold values

	GIS aggregation method	
Optional grid sizes	mame	meme
option 100 100 x 100 m = 1 ha	100_mame	100_meme
option 500 500 x 500 m = 25 ha (average field size of region)	500_mame	500_meme
option land unit	FL_mame	FL_meme

Source: own presentation

The aggregation method can therefore be a crucial point if GIS data is used for handing out payments for soil conservation programmes. For the further analysis in this study, a selection from these combinations had to be made because the use of all the combinations would be far too complex. Therefore, the option of a 100 x 100 meter grid with a mean aggregation method (100_meme) was chosen, since it provides the best available resolution of the three grid size options and reflects sufficiently the soil erosion risk based on the average of the aggregated cells. Even though the mean value method does cause some levelling of the values, the results are still easier to interpret compared to the combination of maximum and average values.

5.2 Erosion risk assessment of cropping practices with a fuzzy-logic model

It was stated in the introduction of chapter 5 that a parameter that describes the erosivity of each agricultural practice on a specific site serving as an indicator in a bio-economic model was needed. Within the cropping practice database of the later described model MODAM, every single work step of a cropping practice is described (ploughing, spraying of pesticides, harvesting etc.). This information can be used to generate such a cropping practice specific parameter.

Because the management decisions (i.e. the chosen crops and tillage practices) have a high influence on the resulting erosion risk, the cropping factor (C-factor) within the USLE (Figure 7, p.59) is calculated more specifically. In this step, the crop and management specific C-factors are derived with a fuzzy-logic tool[10] (Sattler 2007). For each soil category combined with a specific cropping system, a soil erosion risk value is calculated. This provides a comprehensive assessment of the soil erosion risk for all site categories combined with all available cropping practices. The conservation practice factor (P-factor) is set as a fixed value, since there is no empirical data on site-specific conservation efforts in the region.

Table 9 and Table 10 give an overview of the crop types and the corresponding cropping practices that are implemented in the model. For every crop, the possible cropping practices are described in the database. A cropping practice is generally defined as a combination of

- the tillage system (plough or reduced),
- whether intercrops or undersown crops are used,
- the type of fertilizing and
- whether by-products are harvested.

Cropping practices that are not feasible for a certain crop based on expert knowledge are not included in the database (see Sattler 2007).

---

[10] Note that this fuzzy logic model is subject of another research project and was not developed by the author. The approach and all results generated concerning the soil erosion risk properties of cropping practices belong to the works of Sattler (2007), but are confirmed to be used in this study.

The factors used for the evaluation of cropping activities are as follows (Sattler 2007):

a) Water erosion in summer half-year
- crop depending soil cover in summer half-year
- type of cropping practice (plough, reduced tillage, integration of intercrops and undersown crops)

b) Water erosion in winter half-year
- crop depending soil cover in winter half-year
- number of cross-overs during winter half-year
- type of tillage during winter half-year depending on the grown crop
- date of sowing for winter cereals (winter wheat, winter rye)

This information is used within the fuzzy logic expert system, which provides the possibility of ranking the cropping activities according to their erosivity risk. In the second step, the dimensionless erosion risk values of the fuzzy logic expert system were calibrated to the values of known standard practices of each cropping activity (Sattler 2007). This allows the generation of approximated values even for cropping practices that had not been tested in field experiments. Furthermore, this procedure provides a consistent evaluation system for all cropping practices that have not been tested under the same conditions.

All factors are introduced in a step wise procedure into the fuzzy-model that finally calculates a single value for each cropping activity, which represents approximately a cropping practice specific C-factor. In combination with the site specific erodibility values, a potential soil erosion risk in tons per hectare and year can be generated that is similar to the result from a standard USLE calculation (see the following chapter).

Chapter 5 – Soil erosion risk assessment of sites and cropping practices

Table 9: Combinations of cropping practices implemented in the farm model

Key	Tillage	Intercrops	Undersown	Manuring	Harvest by-product
00	Conventional Tillage	no	no	no	no
01	Conventional Tillage	no	no	no	yes
02	Conventional Tillage	no	no	liquid	no
03	Conventional Tillage	no	no	liquid	yes
04	Conventional Tillage	no	no	solid	no
05	Conventional Tillage	no	no	solid	yes
06	Reduced Tillage	no	no	no	no
07	Reduced Tillage	no	no	no	yes
08	Reduced Tillage	no	no	liquid	no
10	Reduced Tillage	no	no	solid	no
12	Conventional Tillage	no	yes	no	no
13	Conventional Tillage	no	yes	no	yes
18	Reduced Tillage	no	yes	no	no
24	Conventional Tillage	yes	no	no	no
25	Conventional Tillage	yes	no	no	yes
26	Conventional Tillage	yes	no	liquid	no
27	Conventional Tillage	yes	no	liquid	yes
28	Conventional Tillage	yes	no	solid	no
29	Conventional Tillage	yes	no	solid	yes
48	Conventional Tillage	(yes)	no	no	no
54	Reduced Tillage	(yes)	no	no	no

Source: Zander 2003

Table 10: Crop types implemented in the farm model

	Cropping practice																				
Crop type	00	01	02	03	04	05	06	07	08	10	12	13	18	24	25	26	27	28	29	48	54
Uncropped arable land							x														x
Peas	x													x					x		
Yellow lupin, grain	x													x					x		
Oats	x	x					x	x			x	x		x	x						
Alfalfa	x																				
Linseed/flax	x													x							
Sunflowers	x													x							
Spring barley	x	x					x	x			x			x	x						
Potatoes	x			x										x							
Corn, silage	x		x		x				x	x						x					
Set-aside	x												x							x	
Spring wheat	x	x																			
Triticale	x	x					x	x													
Winter barley	x	x	x	x			x	x		x											
Rapeseed	x					x															
Winter rye	x	x	x	x	x	x	x	x		x											
Winter wheat	x	x	x	x	x	x	x	x													
Sugar-beets	x	x	x	x	x	x	x							x	x	x	x	x	x		

Source: Zander 2003

5.3 Combination of erosivity and erodibility values

In the last step, the erosion risk value of a cropping activity (fuzzy calculated C-factor) is combined with the site specific erosion risk value (see Figure 11). This procedure ensures that each cropping activity is assigned with a specific soil erosion level on a specific soil type grid cell. A combination of a highly erosive cropping practice with highly erodible soils will result in a high erosion risk. Combinations of high erodibility with low erosive practice will only cause medium levels of erosion risk. Note that the resulting values are based on a cardinal scale, i.e. each combination has one specific value (tons/ha/year). The final evaluation table holds information for the combination of each cropping practice with each erosion risk–soil quality type. Given the number of crops, tillage systems and combinations of intercrops, a total number of 315 cropping activities were evaluated (see Sattler 2007), which were differentiated for the 18 soil categories for soil quality and erodibility.

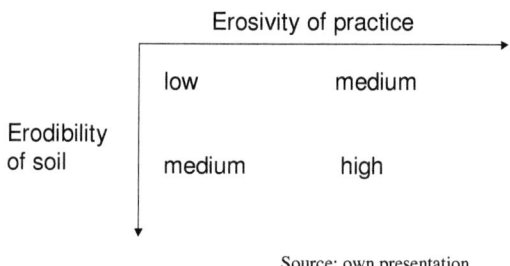

Source: own presentation

Figure 11: Schematic representation of the combination of erosivity (C-factor) and erodibility (soil properties)

An example is shown in growing sugar beets using plough tillage without intercrops on a certain soil type combined with a slope and climate attribute. The same cropping activity showed a growing soil erosion risk (ton/ha/a) depending on the site specific soil erosion risk category, which was based on soil types and topographic factors (see Table 11), which ranged between 0 and 3.60 tons/ha/a.

Chapter 5 – Soil erosion risk assessment of sites and cropping practices

Table 11: Potential soil erosion risk for sugar beets depending on the soil erosion risk category (standard practice: plough, no intercrop, no manure), soil type 38

Erosion risk–soil quality types	Potential soil erosion risk (tons/ha/a)
38_1	0
38_2	0.06
38_3	0.18
38_4	0.57
38_5	1.40
38_6	3.60

Dataset: ZRU1100a AzKl 38 100 grid mean value (Sattler 2007)

Figure 12 shows the example of different crops grown with a comparable standard tillage system (plough, no intercrops) on potential soil erosion rates. Corn showed the highest soil erosion risk (4.2 tons/ha/a), while set aside and alfalfa showed the lowest soil loss (<0.1 tons/ha/a).

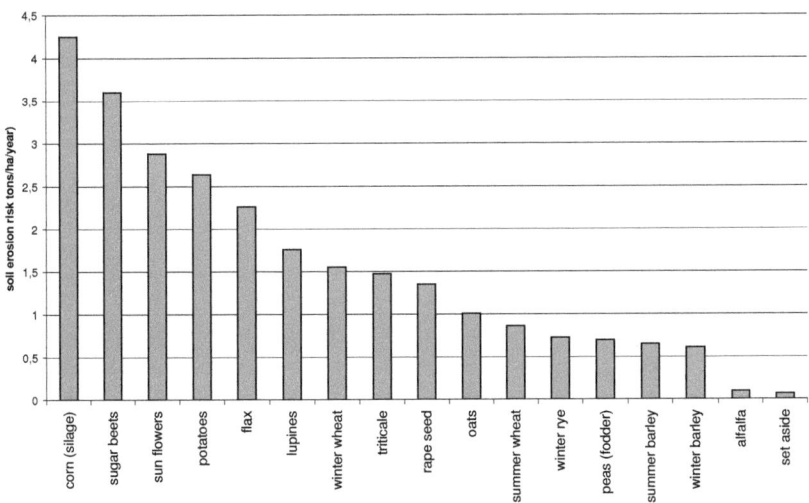

Source: Sattler 2007; own presentation

Figure 12: Potential soil erosion rates of crops with comparable tillage system (plough, standard practice) on sites with the highest erosion risk (dataset: CF38_6, 100meme)

Figure 13 shows the mean value of each crop as well as the maximum and minimum value of each crop. The figure demonstrates that crops with a high mean value (e.g. corn or sugar beets) cannot achieve values of low erosion crops, even when a conservation practice that provides the lowest erosion value for this crop is used. However, it also shows the bandwidth that is given for each crop through the change of the cropping practice.

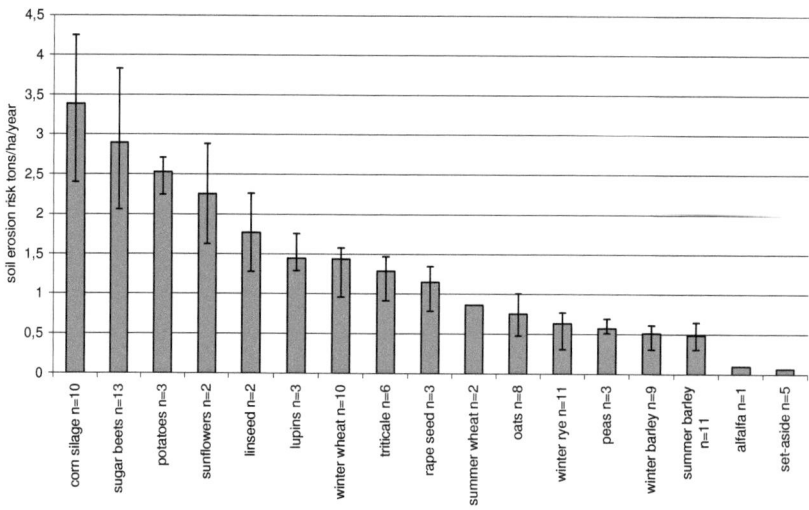

Source: Sattler 2007; own presentation

Figure 13: Mean of soil erosion risk per crop including minimum and maximum value (n= numbers of cropping practices per crop and soil category type)

Discussion of the fuzzy assessment method

A critical point of the fuzzy logic assessment method is (similar to all expert based evaluation systems) that the results are only partially based on empirically gathered data and derived from the experts' rules and knowledge. The values can only be as good as provided by the experts that proposed the way each single work step should be evaluated. However, this approach reduces time and money spent on data gathering. The results are, at least for the standard cropping practices, calibrated on empirically derived C-factors and provide an easy to use database for a consistent ranking of the erosion risk of cropping activities (for more details see Sattler (2007). This approach bears an advantage over pure empirical models, as it allows the expansion of the scope of evaluated cropping practices within a consistent framework without the need for further field trials.

5.4 The effects of erosion risk thresholds in soil conservation programmes

This chapter gives an example of how different thresholds and aggregation options influence the size of eligible areas for soil conservation programmes. Three possible thresholds that can be used to select eligible land for soil conservation programmes will be described. Then, an example of how these thresholds are applied for a specific soil category derived in chapter 5.1.3 is shown.

5.4.1 Threshold options for soil conservation programmes

Firstly, some references from soil science regarding possible thresholds of tolerable soil erosion rates are shown. In soil science, the tolerable soil erosion rate is defined as the amount of eroded soil per year that is below or equal to the regeneration rate, which is the annual amount of new fertile top soil produced from underlying soil material (Schachtschabel et al. 1992). This threshold is based on a strong sustainability criterion. Other thresholds based on political decisions might not meet such strong sustainability criteria. However, they have to be taken into account as the outcome of agreements or compromises between long-term sustainability and short-term decisions of farmers who have to consider competitiveness. For this study, three different thresholds were analysed with two of them based on scientifically derived thresholds given in the literature (Frielinghaus 1998; Schwertmann et al. 1987) and one as a possible agreement on the political level.

5.4.1.1 Zero ton erosion threshold

Frielinghaus et al. (1998) argued that for most soils the regeneration or new build-up of the top soil is too small to be measured. Therefore, they proposed a way of agricultural practice that avoids any erosion from agricultural fields.

This can only be achieved through measures that go beyond simple changes in cultivation options like i.e. reduced tillage. The approach involves an individual screening or the audit of farms with specific consideration for the erosivity of each field. The proposed measures comprise also set aside and the change from arable to grassland for highly erosive parts of a field.

In order to describe the aim of this approach accurately, the resulting tolerable erosion of any cropping activity is set at a threshold of less than one ton/ha/a for any field type. A threshold of zero would result in a non-feasible solution for a farm model, for any agricultural practice would result in at least some erosion (see Figure 13 p.70). Even though the authors tried generally to avoid discussions about thresholds (Frielinghaus et al. 1998), the proposed measures were aimed at levels of almost zero tons in erosion.

5.4.1.2 Variable soil erosion threshold that is dependent on soil quality

Schwertmann et al. (1987) proposed a tolerable erosion level that was dependent on soil quality. Soils of lower quality (e.g. sandy or low organic matter) are less able to tolerate erosion compared to loamy, more fertile soils. The authors developed a simple equation based on the German soil quality index (*Ackerzahl*) ranging from 0 to 100, with 100 representing the most fertile soils in Germany. This index divided by 8 represents a soil quality based threshold for tolerable soil erosion per ha and year (TSE).

$$TSE/ha/a = SQI / 8$$

TSE = Tolerable soil erosion by water
SQI = Soil quality index

For this study, the tolerable soil loss results in the following calculated numbers, see Table 12.

Table 12: Soil quality index classes used in the model and the respective tolerable soil erosion by water

Soil quality index (SQI)	Tolerable soil erosion by water (TSE) (t/ha/a)
25	0.3
38	1
50	4

Source: Schwertmann et al. 1987

5.4.1.3 A Standard Soil Erosion Threshold

Standard-based thresholds, which are discussed on the policy level more than promoted by the scientific world are tolerable erosion rates for specific regions issued by soil conservation authorities. Examples of these thresholds are values of maximum 7.4 tons/ha/year for Ontario, Canada (Stone 2000) or 10 tons/ha/year (Bayerische Landesanstalt für Landwirtschaft (LfL) 2004). Even though such high thresholds are questioned in soil science (Frielinghaus et al. 1998), a standard threshold was used in the model to provide a value that might be of interest in a political process. Calculations were done with an eight ton threshold per hectare and year, representing the highest category chosen for the classification of water erosion risk in Brandenburg by Deumlich et al. (1996).

This procedure was chosen because thresholds provided by scientists were usually questioned and not directly transferred into binding laws during legislation processes. This value can be seen as a scenario for the outcome of a policy making process that mostly does not reflect the most preferred solution from one part of the involved stakeholders.

5.4.2 Applying erosion thresholds as a eligibility criteria

Table 13 summarizes the threshold levels and their references. All thresholds are based on comprehensible criteria from a scientific or political viewpoint. However, in a political process, sustainability is only one criterion that influences decision-making. Therefore, none of these thresholds can be designated as the "true" value.

Chapter 5 – Soil erosion risk assessment of sites and cropping practices

Table 13: Overview for different threshold levels for soil erosion and references

Tolerable soil erosion (TSE) <	1t/ha*a	Soil quality index / 8 (t/ha*a)	8t/ha*a
Notes	Soil recovery or regeneration not measurable under normal conditions (< 1t/ha*a), threshold < 1 t/ha	Tolerable soil run-off in ton/ha*a derived from formula: TSE/ha*a = SQI / 8	a probable compromise, not sustainable, but still a limiting threshold on highly erodible fields
References	(Frielinghaus et al. 1998)	(Schwertmann et al. 1987), p.12)	e.g. (Bayerische Landesanstalt für Landwirtschaft (LfL) 2004)

If these thresholds are applied to the soil categories in chapter 5.1.3, which are based on a standard C-factor, the resulting area above such thresholds could vary extremely. Figure 14 shows the share of "above threshold" soil category within the "100_meme" dataset. If such thresholds were to be used to select eligible land for soil conservation programmes, the share could vary between 8 percent (8ton/ha/a threshold) to 79 percent (1 ton/ha/a) of the total area.

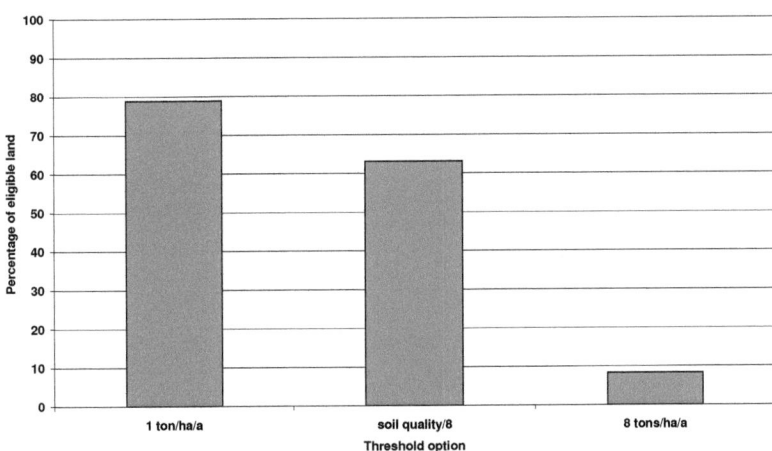

Source: own calculations

Figure 14: Percentage of possible eligible land for soil conservation programmes depending on the threshold levels

If a soil conservation programme is tied to a threshold level, the choice of the selection method is important for the size of the area that the programme is eligible for (Table 14). When a threshold of 1 tons/ha/a is chosen, at least 78.8 percent of the total area would be selected. If the 'mame' method is used for aggregating the grid cells, at least 31.2 percent of the area would be above the threshold level, but the area could rise up to 100 percent.

Table 14: Area with erosion above threshold depending on grid size, aggregation method and threshold levels

Grid size	GIS aggregation method	threshold	area above threshold %
FL	mame	variable	100.0
FL	mame	<1	100.0
FL	meme	<1	100.0
500	mame	<1	99.8
500	mame	variable	99.1
100	mame	<1	95.8
500	meme	<1	94.5
FL	meme	variable	94.3
500	mame	8	86.8
100	mame	variable	84.6
100	meme	<1	78.8
500	meme	variable	70.8
100	meme	variable	63.1
FL	mame	8	34.8
100	mame	8	31.2
100	meme	8	8.2
500	meme	8	5.0
FL	meme	8	1.1

Source: own calculations

The aggregation method, grid size and thresholds can therefore be crucial points, when GIS data is used to hand out payments for soil conservation programmes. By influencing the political process of threshold settings and area selection, stakeholders can vary the size of eligible land and the amount of possible payments per hectare in case such an implementation instrument was chosen.

In order to reduce the computational efforts, only one GIS-dataset for the soil erosion risk assessment was chosen for the bio-economic model, which was derived in chapter 5.1.3. This chapter showed that the analysis of GIS data and the implementation of thresholds bear problems that need further research, but are beyond the scope of this study.

The thresholds described in this chapter will serve as a benchmark for the modelling results of soil conservation policy options in chapter 7. They were not used directly as a selection criteria (target values) for certain crops or cropping practices because such a criteria would be too restrictive in a spatially detailed model.

6 A bio-economic model for the analysis of policy instruments and on-farm measures

6.1 Review of economic models in the context of soil conservation

6.1.1 Different ways of analysis

Soil conservation has been approached economically in different ways. Furtan and Hosseini (2003) described attempts "from quantifying the national impacts of soil loss to identifying the factors which influence a farmer's soil management decisions."

Some studies follow a clear CBA structure, (e.g. Abelson 1979), whereas others use dynamic modelling approaches in order to generate opportunity costs for CBA, (e.g. Lu and Stocking 1998). All approaches use more or less explicitly a utility function either defined as society's welfare (e.g. Abelson 1979) or the individual's profit (e.g. McConnell 1983).

Most of the research activities were done in the inter-temporal modelling of soil erosion because the impacts of soil loss are an inter-temporal problem (Furtan and Hosseini 2003).

Generally, such models employ the optimal control theory of dynamic programming. These models allow decision makers (e.g. farmers) to make investment decisions with a definite planning horizon in mind. Information regarding land tenure, land quality, etc. can also be included in such models (Furtan and Hosseini 2003). Programming models usually try to simulate an individual behaviour of profit maximization, however they also can be used to optimize a social decision making process (Yang et al. 2003). Such models are used as one of the main methods for deriving the *opportunity costs* of specific conservation strategies. These models are used to derive an economic preferable solution for resource conservation. They can show the interdependencies between resource use and the economic performance of an individual or a region in different scenarios; e.g. soil erosion and the economic performance of a farm. This method can show trade-offs between resource use and commodity production that can be used as an efficiency scale. It also can also provide economic solutions based on the optimization of the resource use.

6.1.2 Available modelling approaches

There are numerous possibilities for modelling the economic behaviour of a region or a farm. One of the fundamental questions is whether an econometric model (Forster 2000; Forster and Rausch 2002) or models that follow a programming approach should be applied (e.g. Fox et al. 1995; Yang et al. 2003).

Given the fact that this study evaluates alternative policy designs and scenarios that describe the use of new agricultural practices, an approach is needed that allows for an ex-ante-analysis.

The results and behaviour of econometric models are usually based on reactions that have been observed in the past. Through the use of statistical tools the possible reactions to the change of exogenous variables are estimated. However, for this study the available empirical data was too sparse to base an econometric model on. Furthermore, the effort to get information on the production practices of every individual farmer is too costly. In addition, new instruments had been introduced and different sets of eligible land for soil conservation programmes were chosen, which have not been evaluated before in this context.

Programming models are more suitable for analysing the economic behaviour of farms or regions because of the models' inherent assumption of profit maximization by a farmer. This helps in analysing the adjustments of the model to any given scenario. Programming models are based on observed data and allow for an ex-ante-evaluation (Paris 1991).

Linear programming models and other non-linear programming models had been used to optimize the solution of manifold planning problems for more than 50 years (Paris 1991). The simplex-algorithm developed by Dantzig (1963) is still being successfully used in many applications of this type. The fundamental idea in these models is still the same even though there had been some changes and improvements in the underlying functions and the software used (Arriaza and Gomez-Limon 2003; Umstätter 1999). One of the main improvements in this modelling approach is the improved calibration of these models using empirical data of a base period (Positive Mathematical Programming: PMP) (Howitt 1995; Paris and Howitt 1998). LP models are used both to optimize farm enterprises and to simulate policy effects in support of decision making on the administrative and legislative levels (Heckelei 2002).

6.1.3 An overview of policy relevant studies

In the following, a selection of policy relevant models is presented. Some of the models use an optimization approach that ensures the socially optimal distribution of either public funds or the optimal selection of eligible land for conservation programmes (Fox et al. 1995; Yang et al. 2003; Yang et al. 2005a; Yang and Weersink 2004), while other studies analyse the efficiency of conservation programmes on the basis of empirical data.

The model developed by Yang et al. (2003), designed to show cost-efficient ways for land retirement programmes, assumes a social planner that optimizes conservation policies in a watershed. However, it does not focus on optimal soil erosion rates in the sense of Welfare Economics but on a static optimal distribution of eligible land for conservation programmes. It focuses on how existing pollution rates of a water body can optimally be reduced to a politically set goal achievement.

The advantage of their model is that it can consider the spatial heterogeneity of the land's quality in each watershed, as well as flow paths between the land parcels, but the land use options are restricted due to computational limits. However, from the viewpoint of an SMS-approach, the pollution rates are seen as a given level. The model is used to find the optimal spatial allocation of the conservation programme.

In an earlier study, Lintner and Weersink (1999) developed a model that combined economic, environmental and spatial analysis to examine policies for reducing nitrogen and phosphorus runoff. In their model, the agricultural land use of a whole watershed was optimized in terms of the flow of sediments into a river. The special feature of the model is that the transport coefficients from one cell of the system to the following are endogenous, i.e. the amount of sediment caused by the land use of farm A is known and kept in a cell and the amount that flows over to the next cell, probably owned by farm B is also known. The authors pointed out that this spatial set up bore a positive externality produced by the farms closer to the river, for their fields collected sediments produced by farms up the slope that had more profitable but also erosive practices. "Optimizing management choices and consequently endogenizing the transport coefficients, for all firms simultaneously remove the externality. An empirical application combines hydrological simulation models with an economic optimization model for nutrient pollution of surface and ground water within an agricultural watershed. Although firms are homogeneous in abatement costs, differences in spatial location leave uniform instruments unable to achieve the water quality goal efficiently (Lintner and Weersink 1999)". It is shown, that an ambient tax/subsidy scheme (see Segerson 1988) is more efficient in achieving better water quality. However, "the informational requirements will be excessive in most situations, where the transport mechanisms for residuals are dependent upon the practices of independent decision making units" (Lintner and Weersink 1999), which is usually the case under real farming conditions. The approach is capable of simulating the spatial characteristics of a pollution problem, which results in conclusions on a preferable policy instrument.

Dabbert et al. (1999) developed a model on the landscape level that contained among other indicators, a soil erosion model as well. The economic effects of different scenarios are simulated with a Positive Mathematical Programming approach.

From an analytical perspective, Hediger (2003) showed how a cost-effective scheme for the control of Phosphorous (P) runoff caused by soil erosion from agricultural land into a lake can be set up within a dynamic optimization problem. He used an inter-temporal allocation model with on-site measures to control soil erosion and the resulting P runoff from heterogeneous agricultural land, and the lake-internal measures of water quality improvement. The model was an example for optimal control models, since it assumed complete information for the relation between production and pollution, and the possibility to trace where the phosphorus in the lake had originated. Given the uncertainty of such processes, the model is not applicable for soil degradation as a threat to long-term fertility.

An example of an efficiency survey was given by Forster (2000), who analysed all soil and water protection programs for an entire water catchment area in North America based on a cost-effectiveness ratio (costs per ton of soil saved). In another paper, Forster and Rausch (2002) described the efficiency of government programs aimed at encouraging Lake Erie basin farmers to adopt practices that reduce water pollution. The evaluation was based on the cost effectiveness of program expenditures (i.e., cost per metric ton of soil saved). They found that the majority of Agricultural Conservation Program (ACP) funds appeared to have been spent on less cost-effective practices.

Nakao and Sohngen (2000) evaluated riparian buffer strips by exploring the relationship between buffer size, drainage area size, and effectiveness. Shankar et al. (2000) showed in the example of preventing nitrate discharges that sometimes win-win situations can arise, where both the water quality and the farmers' income increased. Krayl (1993) submitted a study analysing measure-instrument combinations using the example of nitrogen leaching.

In summarizing the aforementioned approaches, it can be said that they are either based on optimization models with partially rough assumptions of the erosive properties of crop practices or crop rotations, but good possibilities for creating scenario calculations for policy design, or the studies use a more empirical approach in order to analyse ex post the efficiency of conservation programmes. The objectives of the studies are either to find the optimal size of conservation programmes in terms of area covered or to find the optimal strategy for soil erosion prevention.

From the viewpoint of this study, an approach that can finds a cost-effective policy that does not rely on or try to calculate an optimal erosion rate should be found. Moreover, the potentials of programming models should be used to simulate the behaviour of profit maximizing farmers, while being confronted with optional policies that provide a safe level of soil use.

6.2 The chosen model system MODAM

The modelling system MODAM (Multi-Objective Decision support tool for Agro ecosystem Management) (Zander 2003; Zander and Kächele 1999) was chosen for this study because it met the needs of this study best. The following chapter describes briefly the structure of the model and shows how the economic analysis is done with a Linear-Programming model that simulates the reactions of a regional farm towards different frame conditions. The data structure is described later in chapter 6.4. For further details regarding the model approach, see (Zander 2003; Kächele 1999).

The different parts of the assessment, as well as the activities and constraints are brought together in this model (see Figure 15):

- the soil conditions are classed using an adapted version of the Universal Soil Loss Equation (USLE),

Chapter 6 – A bio-economic model for the analysis of policy instruments and on-farm measures

- the environmental evaluation for the cropping activities being based on the fuzzy-logic tool,
- descriptions of plant and livestock production activities,
- the policy framework comprising EU-payments and regulations,
- the farm assets (e.g. land, labour, machinery),
- and prices.

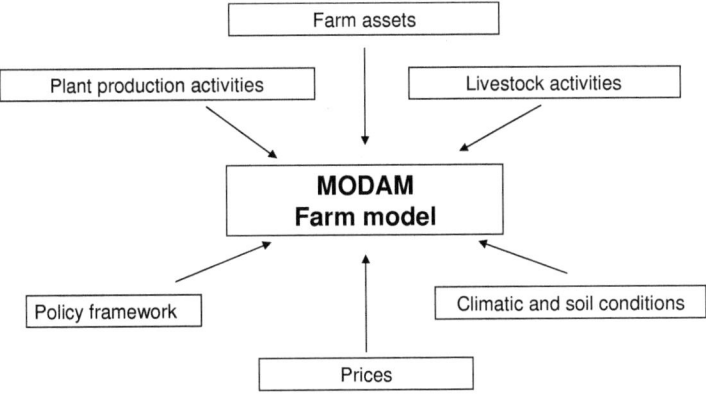

Source: own presentation

Figure 15: Data types used in the MODAM farm model

Exogenous variables in this model are the soil conditions, the descriptions of plant and livestock production activities, the policy framework, the farm assets and in- and output prices. Endogenously calculated variables are the environmental and economic assessments of the cropping practices (soil erosion risk and gross margins).

The on-farm costs of erosion preventing measures are one of the crucial points when governmental administrations try to develop soil conservation policies. On the one hand, doing field trials is time consuming and expensive, on the other hand, individual farm data from case studies are often biased and not representative (Vereijken 2001). The objective of this model is to generate with acceptable effort, on-farm costs using average or prototype data from farming activities.

The modelling system MODAM is appropriate for meeting such needs. MODAM (Zander and Kächele 1999) comprises two parts: a bundle of large databases, describing the regional agricultural practices in great detail and a linear programming tool to simulate decision behaviour when farmers produce economically under the conditions of soil conservation policies. The tool can be used for evaluating agro-economic scenarios with respect to their regional effects. For a list of further applications of this approach see (Zander 2003).

Furthermore, the underlying ACCESS-database system allows one to work with large, highly detailed datasets and generates much bigger matrices than in the case of the usual spreadsheet-tables. The exchange between other software such as fuzzy tools for the evaluation of the environmental impact is facilitated.

In order to outline the on-farm costs and the environmental effects of soil conservation measures at the field level, the agricultural practices need to be described in the model in a very detailed way. The model itself consists of hierarchically linked modules, which are grouped into three main steps (Figure 16). Step 1 describes the farm or region with its production capacities and activities. In step 2, a partial evaluation of the economic and ecological effects is performed (i.e. the gross margin and ecological evaluation of agricultural activities). In step 3, the economic behaviour of the farmer is simulated by a linear programming module, which ensures that production factors are allocated according to their best factor utilisation.

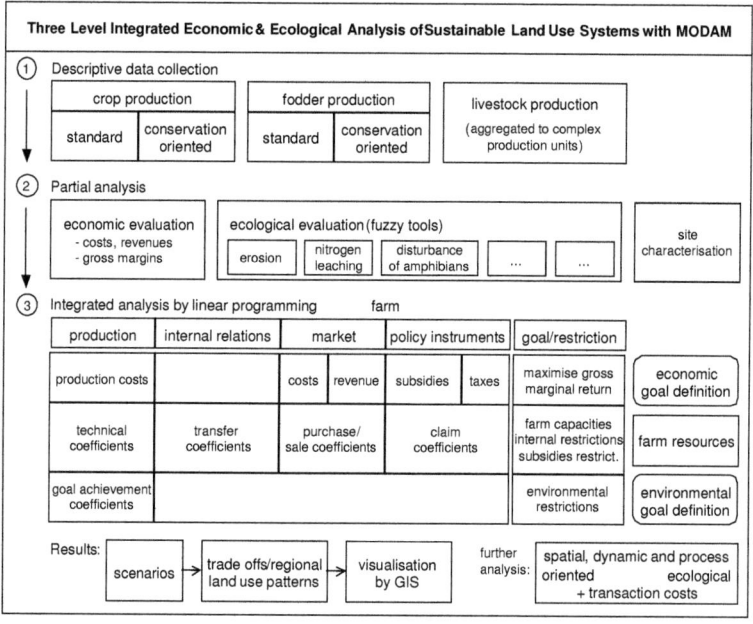

Source: Zander 2001

Figure 16: Three level, integrated economic and environmental analysis of agricultural land use systems with MODAM

Step 1:

Information on the farming activities is based both on expert knowledge and interviews. The production factors (e.g. labour) correspond to actual situations or are adjusted according to available statistical data. The descriptions of standard and adapted cropping practices are based on expert knowledge resulting from research at the Centre for Agricultural Landscape and Land Use Research

(ZALF) (Meyer-Aurich et al. 1998; Zander 2003). Livestock practices are formulated on the basis of standard data tables (KTBL 2000).

These sets provide information on the on-farm demand and supply of dairy, pork, chicken and sheep production, which are already assembled for livestock farm types at different levels of productivity (Kächele 1999). (The used data for the regional model will be further described in chapter 6.3.)

In the MODAM database, every measure or work step in plant and in livestock production is described in great detail. This provides information about any specific pesticide application, fertiliser usage or the time periods when a work step is done on a field. This information may be used for further ecological evaluation of environmental objectives, which is one of the main differences from other models and databases used for the assessment of agro-environmental issues (Roedenbeck 2004).

For several production intensity levels, either for crop or animal production, a specific set of input combinations is generated for each product, corresponding to different points on a hypothetical production function. For a list of the contained crops and cropping practices see Table 9 and Table 10.

Step 2:

The economic assessment of the crop, forage and animal production practices takes place in a separate database. Costs are calculated with respect to the farm machinery, interest costs and price levels, as defined for the specific scenario. Economic yields from all sold outputs are based on the scenario-specific prices. Energy consumption and required labour per management operation are derived from standard data tables (KTBL 2000). The results are summarised in detailed tables characterizing each production technique with different cost categories, its yields and its gross margin. The environmental assessment of agricultural practices can be based on either relatively simple, statistical tools (see Meyer-Aurich 2001) or on more detailed fuzzy tools, but both are based on expert knowledge. For this study the fuzzy logic approach was chosen (see Sattler 2007 and chapter 5.2), so that it was possible to assign values of environmental performance (i.e. soil erosion potential) directly to cropping practices.

The value of each practice for a specific environmental good is expressed by continuous assessment values. Here, it is expressed as a value of potential soil erosion risk in tons per hectare, describing the potential erosion effect of a specific cropping activity on a specific type of soil (see chapter 5.2), see also (Zander 2001).

Step 3:

The behaviour of farmers must be described so that the model can incorporate the effect of regulatory conditions like subsidies or regulations. This is done through a Linear Programming tool with the assumption that farmers' profit maximizations are subject to certain restrictions. Although farmers obviously have objectives other than profit maximization, these are neglected for reason of simplicity. It is assumed that most of farmers' decisions are based on economic rationality (Kächele 1999, p. 4). The optimization process simulates the farmer's decision in the production of a set of possible crops and livestock with given prices, labour force and field sizes. Crop rotation effects are described by the share of each crop grown on one field type (more details in chapter 6.3). Although most applications of the model use a static version, the model can also be run as a dynamic recursive version (Zander 2003).

In a general way, the model can be described as follows (Paris 1991):

$$\text{Max TGM} = \sum_{j=1}^{n} c_j x_j$$

$$\text{subject to} \quad \sum_{j=1}^{n} a_{ij} x_j \leq b_i, \qquad i = 1,..,m$$

$$x_j \geq 0, \qquad j = 1,...,n$$

with:

    TGM = total gross margin
    c_j = unit gross margin of commodity j
    a_{ij} = amount of input i necessary for the production of one unit of commodity j
    b_i = total available quantity of input i
    x_j = quantity of commodity j produced using the technological process j

The soil conservation effect is introduced to this equation as the "production of environmental goods or bads", in the case of soil erosion the "production of soil erosion". The production of each commodity j is linked to the production of the environmental commodity k_j. This production of k is inserted into a restriction, which can be interpreted as a minimum environmental threshold or standard. This standard could, for instance be defined by reference to a "good technical practice" or a safe minimum standard. This value can be parameterized to build scenarios of increasing demand for environmental goods like soil conservation. The activity's value per field type can also be used as an indicator of the environmental performance in this specific area.

This additional environmental constraint can be written as follows:

$$\text{subject to} \quad \sum_{j=1}^{n} d_{kj} x_j \geq e_k, \qquad k = 1,..,o$$

    d_{kj} = amount of output k associated to the production of one unit of commodity j
    e_k = minimum amount of environmental quality k

This soil erosion restriction in the model can either be set for the whole region or specifically for each soil and erosion category. This allows one to analyse the region as one regulation area or to focus on targeted areas with a higher erosion risk.[11]

If the model is run in a parameterized version, the results can be visualised in the form of trade-off functions between different environmental and economic goals. Also, abatement cost functions can be derived.

In order to outline the model briefly, the main characteristics are again summarized as follows:

1. The MODAM approach is based on a linear programming solver that optimizes the total gross margin of a farm or a region
2. The calculated scenarios in this study are comparative-static, dynamic effects are neglected.
3. The agricultural activities need to be described in high detail (i.e. each work step is described and includes time span, machinery used, labour time) in order to allow for an ecological evaluation (i.e. soil erosion risk).
4. The model system consists of hierarchically linked databases that facilitates the generation of huge LP-matrices.

6.3 A choice between programming models

The following paragraphs will discuss other mathematical programming approaches and compare them to the chosen approach.

When focussing on different mathematical programming models, the question of what specific type of this model family should be used arises. The scope goes from the basic linear programming approach (Dantzig 1963) to the recent positive mathematical programming (PMP) models (Howitt 1995; Röhm 2001). Furthermore, the used functions that define the economic activities can be linear or quadratic. Quadratic functions limit the model's tendency to overspecialize. Some authors also used Cobb-Douglas functions or CES function (Constant elasticity of substitution), but the use of these functions is still limited (Röhm 2001) and therefore will not be discussed further here.

---

[11] With MODAM, it is generally possible to calculate scenarios both for different goal achievement levels (goal driven scenarios, GDS) and for different policy instruments (policy driven scenarios, PDS) (Zander 2001). A GDS would consist of a stepwise variation of the above mentioned environmental restrictions. A PDS can be attained by change of prices and subsidies, the introduction of quotas (change of input or output quotas) and legal interventions changing the possible production practices. In this study, one goal driven scenario is calculated in order to serve as a reference scenario for policy options (assuming optimal control). Furthermore, several policy driven scenarios are analysed, that consist of different procedures of area selection and variation of policy instruments.

In the search for an appropriate model approach, the PMP model type was considered a possible option for this study instead of the standard linear solving algorithm that is used in the described MODAM model system.

Röhm (2001) and Umstätter (1999) discussed the advantages of positive mathematical programming in comparison to the Linear Programming approach. By using the calibrating procedure with additional marginal cost functions in the PMP approach, the typical disadvantage of the standard LP model is eliminated. Standard LP-models tend to skip between extreme solutions after little changes (e.g. in- or output prices) in the model parameters are made, which finally leads to over-specialized model results. In some cases, the standard LP-model is only limited by a few extremely binding restrictions. Usually, calibration restrictions are introduced to fit the model to a base period. By using these restrictions extensively, the model is finally fixed on the base period solution. Even though it is not obvious, a PMP model is also fixed on a basic period, although the approach allows some flexibility towards other solutions.

In the MODAM approach, over-specialization is avoided through the high number of activities competing for the production factors and detailed breakdown of restrictions, e.g. time spans, fodder restrictions. Therefore, the solution space of the analytical problem is much wider than that for a model with only few activities and restrictions.

Models of the PMP type are more or less inflexible when new activities, which have not been used in the basic calibration period need to be introduced in the model (Röhm 2001; Umstätter 1999). The estimated (invisible) high costs that had impeded farmers in the basic solution from using these activities, finally make it very unlikely for the model to use these new activities. As a result, the over-flexibility of the standard LP-model with its unstable solutions is converted into an overly-binding PMP model that only changes among the activities that where used in its basic solution.

This disadvantage is crucial for the choice of model to be constructed in this study: Many of the soil conserving cropping activities had hardly been used so far and, more importantly, relevant data on reduced tillage is not available (KASSA 2006). Usually, only data on the crops grown in a region are available, but it would require too much effort to find in percentage the crop cultivation activity for every variant (e.g. conservation tillage vs. conventional tillage). The result of the model would finally be biased by whether a cropping activity had been used or not, because the additional costs in PMP modelling of a cropping variant depend only upon whether a cropping activity was used or not. Furthermore, finding consistent empirical costs for these practices will be even more difficult. The workarounds to counter these effects usually bring about other disadvantages, which will not be discussed here (further reading: (Röhm 2001; Röhm and Dabbert 2003).

Another disadvantage of the PMP models is the use of shadow prices for calculating the cost coefficients for the calibration functions, which is based on the assumption that shadow prices reflect the actual opportunity costs of each activity (Umstätter 1999). In a low regulated agricultural environment this assumption might be true, but in the case of EU regulated markets with subsidies and production quotas, shadow prices hardly reflect the true price or cost of a certain good or activity.

Considering the above arguments, the advantages from the detailed resolution of the modelling system MODAM simply outweigh the concerns that come with the use of the solver type (PMP). Therefore, the use of MODAM for this study is seen as appropriate. In comparison to other models that evaluate economic and ecological effects, the number of possible cropping activities in MODAM is higher. This allows the model to find solutions among manifold possibilities. The huge number of environmental and farming restrictions contained in the MODAM model creates results that are not as specialized as those produced when a standard LP-model with a low number of activities and restrictions is used.

6.4 Modelling the agricultural sector of the study region

The farms in this region were aggregated into one regional farm, using both the above mentioned empirical data for describing the assets of the region and calibrating the model, and statistical data from representative cropping and livestock activities. The following chapters give an overview of the type of data (modules) used to describe the region and how the data were transferred into LP-structure.

MODAM consists of data bases that generate activities, restrictions, cost variables and transfer variables of the LP-table. If a specific group of variables is not needed (e.g. if a farm does not have livestock), the module can be omitted from a query list or if further restrictions or other parts of the model have to be changed, only parts of the whole LP-table is recalculated, which helps reduce time used (effort) in the model building phase.

The modules used in this model version and some explanations for each group are shown in Table 15.

6.4.1 Costs

The cost module generates the objective function of the LP module. It provides all costs for the farm's activities (be it the costs of plant or livestock production, labour or land costs) and the possible positive yields generated by selling activities. MODAM follows a disaggregated approach, i.e. the optimum solution is not based directly on the activities' gross margins but on the optimization of the variable costs and the possible yields of certain cropping activities. This allows for example, for cereals to be sold or to be used on the farm as fodder.

The entire model searches for the optimal solution by maximizing the total gross margin of the region. Prices are based on 2002 data and are adjusted where appropriate (see following chapters).

Table 15: LP modules used in the model

Module	Description
Costs	Objective function The model is optimized by minimizing the total costs, i.e. maximizing the total gross margin of the region
Land data(Surface of field types)	Provides area for cropping activities and pasture
Labour supply	Provides labour input for both cropping and livestock activities. For the cropping sector, the labour module is divided into time steps, in order to describe time requirements during peak seasons.
Policy conditions for crops	EU payments for crops
Policy conditions for livestock	EU payments for livestock sector, changed according to the EU-Reform
Crop production	Yields, work steps, inputs of cropping and pasture activities
Soil erosion	Soil erosion module describing the effects of cropping activities on specific soil categories
Crop rotations	Crop rotation limitations based on good management practice
Cropping restrictions due to quotas and delivery rights	Manually set restrictions. In the case of the model, potatoes and sugar beets are restricted to the empirically found value, since both crops are only grown when delivery rights exists
Livestock systems incl. feeding	Livestock system with all categories described above (reproduction, labour needs, buildings etc.) Feeding demands from livestock systems and provision of nutrients by cropping and pasture activities
Manure management	Module for the management of supply and uptake of manure

Source: own presentation

6.4.2 Land data

As described before, the agricultural land of the model region was selected from GIS land use data (see Chapter 5.1). The arable land was grouped into three soil quality categories that are based on the German fertility index (dt. *Ackerzahl*). Each 100 x100 meter grid cell of the soil data base was assigned with the aforementioned soil erosion risk value. Finally, all grid cells were subdivided into erosion classes from Table 6. Grasslands were not included in the erosion assessment since only limited erosion is expected on pasture land. As a result, 18 soil types were grouped and 2 grassland types are available in the model.

Table 16 shows the distribution of land within the different categories. Each subgroup describes a separate land supply in the regional model, so that the land use on each soil type can be analysed separately.

Table 16: Surface distribution of different site qualities and erosion classes in the model region; example for a 100 meter grid model (for erosion classes: see Table 6)

Soil quality category	Erosion class	Surface ha	∑ ha
25	1	10	
	2	87	
	3	444	
	4	714	
	5	112	
	6	20	
			1387
38	1	39	
	2	163	
	3	1379	
	4	4168	
	5	1438	
	6	808	
			7995
50	1	35	
	2	106	
	3	724	
	4	2673	
	5	871	
	6	329	
			4738
Grasslands Low quality	-	500	
Grasslands High quality	-	1850	
∑			2350
Total		16470	

Source: Sattler 2007

6.4.3 Labour supply

Labour supply is introduced in the model as an unrestricted resource, which can be hired on an hourly basis. Given the high unemployment rate in the region especially among former farm workers, it can be assumed that skilled farm hand labour force is easily available. Some farms in the region hire part of their employees only during the work intensive periods of the year and release them during winter months. Labour costs are based on the average hourly wage for skilled workers in 2002 (€7.41/h) (Statistisches Bundesamt Deutschland 2006).

6.4.4 Policy conditions

The model was developed based on the Agenda 2000 policy conditions (European Commission 2006; Uthes 2005) (see Table 17), with area payments for specific crops, animal specific subsidies, milk quota, set aside obligations etc. Since the policy change that finally took place in the beginning of 2005, the old policy framework is no longer valid. However, the calibration of the model could only be done using data from before the policy change. The recent reform of the Common Agricultural policy (CAP) (BMVEL - Bundesministerium für Verbraucherschutz 2005; European

Council 2006c) was adapted by using the Agenda 2000 region model data with a different policy framework. The policy framework describes the 2003-CAP reform, which contains decoupled standard area payments.

Table 17: Subsidy levels in Agenda 2000 and the CAP-reform of July 2003

	Agenda 2000 Subsidies and Regulations for Brandenburg	CAP 2013 decoupled payments and estimates for the amount of individual payment per ha in region
Area payments		
Grandes cultures (cereals)	285 €/ha	290 €/ha
Protein crops	328 €/ha	290 €/ha + 55,57 €/ha (coupled share of payment)
Oil seeds	343 €/ha	290 €/ha
Pasture (starting with 98 €/ha in 2005)	-	290 €/ha
Set aside	285 €/ha	290 €/ha (with a slightly higher payment for additional set aside)
Set aside regulations		
Minimum size per application/farm	0,3 ha	0,3 ha
Minimum size per plot to be set aside	0,3 ha	0,1 ha and >10m wide
Obligatory set aside per farm		
minimum	10 %	8,73 %
maximum	33 %	not limited
Livestock payments and regulations		
Special male premium (cattle)	210 €/head (bulls)/ 150 €/head (ox)	All livestock related payments including milk premiums are no longer paid based on production levels. Instead, payments from a reference period (2001-2003) are used to calculate individual payment titles on a per hectare basis that can only be activated when one hectare of agricultural land is used. These payments will decrease by 2013, when the pasture land payment is fully implemented. For this model, a complete decoupled, equal payment is assumed.
Suckler cows incl. Heifers premium	200 €/head	
Extensive livestock husbandry	100 €/head	
Slaughter premium for adult bovine animals	80 €/head	
Calf slaughter premium	50 €/head	
Additional payments cattle slaughter premium	23 €/head	
Milk premium	8,15 €/t	
Additional milk premium	3,66 €/t	

Source: own presentation; adapted from Uthes 2005

This model uses a regional average payment per hectare, which is planned to be realised in 2013. Since the new CAP-reform payments are handed out on a regional level and are no longer coupled to a specific crop production, all former production based payments were deleted from the model (Agenda 2000). Except for a slightly higher premium for protein crops, farmers have no more incentives to orientate their production on the different levels of subsidies. Furthermore, the individually calculated payments based on former livestock production cannot be simulated on a regional level. However, this is not necessary, since the payments too are not related to production.

The level of the subsidy used in the model corresponds to the envisaged standard area payment in Brandenburg for arable and pasture land.

The only condition for receiving one payment unit through the CAP reform is to work on one hectare of agricultural land that had been enrolled in an EU-database. Therefore, these payments can be seen as money transfers to land users or in the long term, transfers to land owners, since most land owners will try to gain back most of the rent that can be generated by their land. Given the decreasing supply of agricultural land due to other land uses, the position of land owners is likely to put them into a position to gain an increasing share of this rent. However, given the static approach of this model, the payments are still included separately. The model cannot reflect this process of rent transfers between land users and land owners. In the long run this could lead to a decrease in profits for land users but it is assumed to have no distorting effect on this study, since the relative preferability is constant for the different production practices.

6.4.5 Crop production

The database in MODAM contains cropping activities for all crops grown in the region. IACS[12] data for the years 2000 and 2001 were available for the federal state of Brandenburg (Matzdorf et al. 2003), aggregated project data), which show the crop shares grown in the region on precise field levels. For this study it could be referred to data that had been aggregated on the municipality level. The crop shares were similar for both years.

Table 18: Crop shares on arable land of study region in 2001 based on aggregated project data

Crop	Share in percent
Winter wheat	32.59
Winter rye	6.08
Winter barley	10.61
Oats	1.06
Corn silage	6.58
Rape seed	15.69
Potatoes	0.01
Triticale	11.02
Peas	0.45
Lupines	0.04
Set aside without non-food crops	4.45
Set aside with non-food crops	4.37
Sugar beets	3.97
Other crops	3.08
Total:	100

Source: Matzdorf et al. 2003

---

[12] Integrated Administration and Control System. IACS is used as EU-wide for the administration of EU area based payments.

The table above shows the numbers for the study region in 2001 (Table 18). In order to validate and calibrate the regional model, the study region crop shares were compared to the crops produced in the status quo model result (see chapter 7.4).

A multitude of cropping practices (see Table 9 and Table 10 p.67) is described for all crops in the database containing the information on work steps and the machinery needed (see Figure 17). This allows for both a detailed economic analysis and a well founded soil erosion risk assessment.

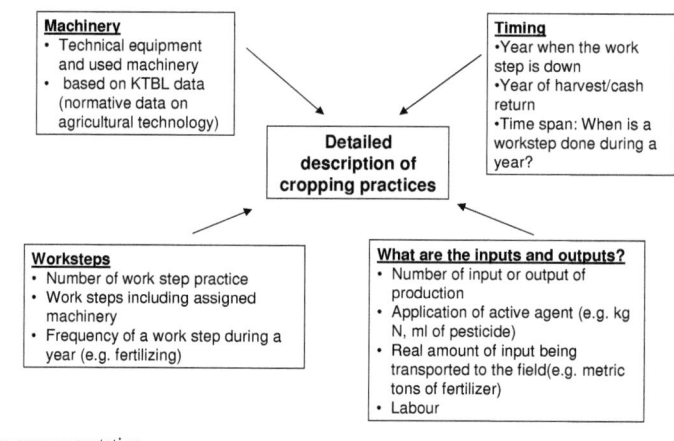

Source: own presentation

Figure 17: Description of cropping practices in the MODAM database

The standard practice is based on tillage with plough, no intercrops or under-sown crops, no organic fertilizer application (manure) and not harvesting by-products such as straw. Yields are based on soil quality and precipitation with precipitation being equal for the whole region (see the following equation).

Crop yield = f (crop, soil quality, cropping practice, preceding crop, precipitation)

Each cropping activity is defined for the three soil quality groups in the study region to reflect the increasing yields that depend on the soil quality. Table 19 gives an example for winter wheat with three assigned yield classes. For more information see Zander (2003).

Table 19: Crop yields for winter wheat depending on soil productivity class

Crop	Soil productivity (AZ)	dt/ha
Winter wheat, standard practice, plough	25	32.3
	38	57.0
	50	71.9

Source: Zander 2003

For the grassland activities, both an intensive and extensive variation are implemented. The model describes activities for hay, silage, grazing and forage based on two site-specific yield expectations (Kächele 1999).

For the Linear Programming matrix, a cropping activity is described with its costs, yield, labour and area requirements, EU area payments, rotation restrictions, manure uptakes and soil erosion properties. Table 20 gives an example of the transfer variables for one cropping activity (winter barley, plough, soil fertility class 38, organic fertiliser used).

Table 20: Example: Variables for the activity "winter barley" in the LP-matrix

Category	Unit	Value
Labour needs (March-May)	Hours per hectare and year	0.12
Labour needs (May-July)	Hours per hectare and year	0.12
Labour needs (July-August)	Hours per hectare and year	1.19
Labour needs (August-September)	Hours per hectare and year	2.63
Labour needs (September-October)	Hours per hectare and year	0.95
Labour needs (October-November)	Hours per hectare and year	0.53
Phyto-sanitary restrictions	Membership term indicating that crop belongs to a specific rotation group	1
Uptake organic manure: Nitrogen	kg	100
Uptake organic manure: Phosphate	kg	50
Uptake organic manure: Potassium	kg	137.5
Yield grains	dt/ha	40.57
Yield straw	dt/ha	10.69
Area demand for one unit of activity	ha	-1
Area payment demand	factor	-1
Area payment restriction (oil seeds)	Percentage of land with crops eligible for payments	-0.08
Set aside maximum	Percentage of land with crops eligible for payments	0.33
Set aside minimum	Percentage of land with crops eligible for payments	0.10
Potential soil erosion	Tons/ha/year	0.22

Activity key: CFWGE1102aCF38_5 Source: own presentation

6.4.6 Soil erosion

The site specific soil erosion parameters from chapter 5.2 were introduced into the LP model in an activity for soil erosion, which allows one to read the total amount of erosion in the region from the calculated value of this variable (see Figure 18). Each cropping activity in a specific soil category delivers its soil erosion value into the erosion restriction row which is fixed to zero. A transfer variable of the value 1 writes the sum of erosion created by the cropping practices into the erosion activity column. As a result, each cropping practice that contributes to the total gross margin of the model (active in the LP-model), is counted with its erosion value in the erosion activity column. This module was also extended to a site specific level: since the erosion coefficients were available for each soil category, it was possible to split the erosion activity into soil categories. For each soil category a specific soil erosion amount is generated. A maximum limit on the erosion activity variable allows for the calculation of scenarios with exogenously reduced levels of soil erosion.

			Erosion activity	Cropping activities		
				wheat	rye	sugar beets
Unit			tons	ha	ha	ha
Activity			13	5	3	2
Erosion restriction	0	=	1	-0.9	-0.3	-3.8
			Transfer variable	Erosion coefficients		

Source: own presentation

Figure 18: Example layout of the LP-erosion module

6.4.7 Crop rotations

MODAM contains phyto-sanitary crop rotation restrictions for all crops that are based on Good Technical Practice and expert knowledge (Zander 2003). For each main group of crops (e.g. cereals, oilseeds) a maximum share is given. Within the main groups, single crops have more specified values. Table 21 shows the values for the main groups and crops used in the study region.

Table 21: Crop rotation restrictions based on Good Technical Practice

Crop	Max. share in rotation in %	
	Main group value	Crop value
Cereals, general	0.75	
Wheat		0.25
Rye		1.00
Barley		0.50
Oats		0.25
Triticale		0.33
Oilseeds	0.50	
Rapeseed, mustard		0.25
Sunflowers		0.20
Linseed, flax		0.20
Root crops	0.50	
Potatoes		0.25
Beets		0.20
Corn		0.50
Legumes (large grain)	0.25	
Peas		0.20
Beans, lupines		0.25
Soy beans		0.25
Fodder legumes (general)	0.50	
Alfalfa		0.33

Source: Zander 2003

6.4.8 Quotas

Since quotas for sugar beets are not known, an overall production quota is assumed for the region based on the average yield on soil group 50 multiplied by the area used in 2002 for growing sugar beets, resulting in a quota of 23320 tons for the study region. This procedure ensures that not more sugar beets are grown than in 2002. For the scenarios under CAP conditions, quotas are no longer distinguished between A- and price reduced B-quotas, therefore prices for sugar beets are at the level of B-quota.

A further restriction was introduced for potatoes which are usually not grown without contracts with processing companies. Therefore, the share for potatoes is also limited to the area found in 2002 (16.47 ha).

In the milk sector a quota was introduced that covers the current yearly milk production of the milking cow numbers.

6.4.9 Livestock systems

The livestock sector in MODAM is described by complete husbandry systems that comprise a detailed description of stable type, milking systems, feeding needs, yields and labour requirements (for a detailed description, see Kächele 1999). Figure 19 gives an overview of the different data types that describe a livestock system. The costs in the different livestock systems that are used in the LP module are derived from this data. Furthermore, a cost optimal feeding plan is calculated in the LP based on the nutrition needs of each animal group in a livestock system and the available feeding crops.

Fodder is supplied through the different pasture and cropping practices. The relevant nutrients that are used for the ration calculation are protein, energy and a minimum share of fibre. A further restriction limits the total amount of fodder that can be consumed per day (dry matter). Nutrient demands per animal are taken from standard feeding tables (DLG 1997, KTBL 1998). A feeding ration is optimized in the LP based on the yields of crops and pasture grown in the region.

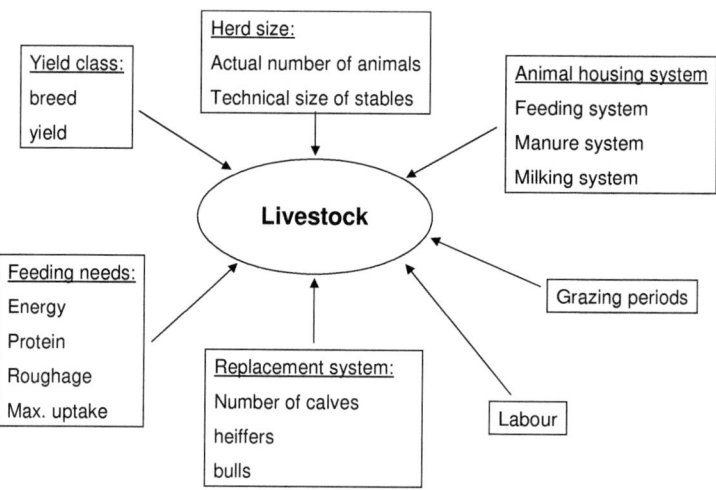

Source: own presentation

Figure 19: Data describing a livestock system in MODAM

Table 22 shows an example for the transfer coefficients of one milking cow in the LP-table (e.g. nutrition demand of a dairy cow, labour, products).

Table 22: LP-coefficients for a dairy cow in the LP-model

Description	unit	value per year	per day
labour time needed	hours per year / min per day	33.95	5.58
protein needed	g	747999.95	2049.31
crude fibre	kg	1144.00	3.13
dry matter	kg	5720.00	15.67
energy (specific for lactating cows)	MJ NEL	34335.98	94.07
manure	kg	6883.74	18.86
stock replacement (0,25 per year)	kg	130.00	
milk production	kg	7000.00	
milk quota needed	kg	7000.00	
offspring male (0,5 calf/year)	kg	21.83	
offspring female (0,5 calf/year)	kg	21.83	
carcass	kg	137.50	
stable need	#	1.00	

Source: KTBL 1998

The livestock number is assumed to have an upper limit in terms of total animals kept but there is no minimum level of livestock production. This is done to simulate the short or medium term reactions of farms that might produce less at lower prices but are not able to produce more at higher prices because of restricting stable capacities (investment activities are not taken into account). The numbers are shown for each livestock type in the following tables.

For the description of the region's livestock capacities, statistical data describing the livestock numbers in the Uckermark administrative district was used (Landesamt für Verbraucherschutz und Landwirtschaft Brandenburg 2003a), see Table 23.

Table 23: Livestock numbers for the district of the model region

Livestock per 100 ha agricultural area as of May 1999										
Administrative district	Horses	Cattle		Pigs		Sheep	Fowl		Geese	Ducks
		total	thereof milking cows	total	thereof sows		total	thereof laying hens 1/2 year and older		
Uckermark	0.6	36.1	11.8	42.0	5.4	9.5	4.2	4.0	0.1	125.1

Source: Landesamt für Verbraucherschutz und Landwirtschaft Brandenburg 2003a

The animal counts per 100 ha of this survey were used to describe the livestock sector of the model by multiplying these stocking rates with the model region's agricultural area (16470 ha agricultural land). This procedure was chosen since the share of agricultural land for the district is similar to the study region's share (see Table 1) with the assumption that stocking rate is evenly distributed over the district.

In order to simplify the model, horses, sheep and poultry were not included in the model. A feeding pig system was used to simulate total swine production and manure quantities. Cattle production was divided into a milking cow and a bull fattening part. The livestock systems used for the model are shown in the tables below.

The dairy sector of the regional model was separated into different age steps (see Table 24). Male calves were assumed to be sold after 4 months. Part of the female calves are used for the replacement of older cows, the other part may be sold as well. The number of animals for the region was adjusted to the empirical value of milking cows (Table 23) using an integer number of model stables (see below) that was added to the model.

The average amount of milk produced per cow was adjusted to 7000 litres/cow/year to correspond to the average of the administrative district (Landesamt für Verbraucherschutz und Landwirtschaft Brandenburg 2003b).

Each model dairy farm consisted of 200 cows including the reproduction needed. In order to reach the region's livestock number ten dairy farms were assumed.

Table 24: Animal groups in a dairy cow system of the model region

Animal	Number of animals per stable	Resulting animals per 100ha	Total number of stables	Total number of animals
Calf female < 4 months	32	1.0	10	320
Calf male < 4 months	32	1.9	10	320
Calf female < 0.5 years	16	1.9	10	160
Heifer; 0.5 to 1 year	48	2.9	10	480
Heifer; 1 year to 18 months	48	2.9	10	480
Heifer; 19 to 27 months.	73	4.4	10	730
Dairy cow	200	12.1	10	2000

Source: own calculations

Table 25: Animal groups in the bull fattening system of the model region; own calculations

Animal	Number of animals per stable	Animals per 100 ha	Total number of stables	Total number of animals
Calf male < 0.5 year	11	1,0	15	165
Bull 0.5 to 1 year	32	2,9	15	480
Bull 1 to 1.5 years	32	2,9	15	480
Bull 1.5 to 2 years	26	2,4	15	390

Source: own calculations

The bull fattening system (see Table 25) was represented by units of stables that produce 50 bulls per year. The system was also separated into four different age steps. 15 stables of this type were needed to meet the region's livestock number for cattle other than dairy cows.

Pork production in the model region was represented by units of 500 pigs per farm (see Table 26), with piglets bought outside of the region. Nutrition demands were based on energy, digestible protein and lysine. These data were also taken from standard nutrition tables (DLG 1991; DLG 1993). An upper limit for total uptake per animal was introduced as a restriction.

Table 26: Hog production system in the model region; own calculations

Animal	Number of animals/stable	Animals number/100 ha	Total number of stables	Total number of animals
Hogs (Feeding pigs)	500	42.5	14	7000

Source: own calculations

6.4.9.1 Costs for stables and labour input

The costs of investments in buildings for livestock were taken from the KTBL data (KTBL 2000). These costs were then divided by the number of animals kept per building. In the LP-matrix it is possible to exchange places for animals of the same livestock system (i.e. within cattle or pigs). This allows for changes in the number of animals if it is economically more preferable to increase numbers in one age class and reduce in another.

The labour needs of the livestock system for feeding, manure management and milking were also taken from the KTBL data (KTBL 2000). All data were adjusted to the livestock system size.

6.4.9.2 Manure management

All livestock in the model are mainly based both on liquid manure and dung systems. Livestock systems deliver nutrients into a manure equation, while manure based cropping activities take nutrients from it. The main plant nutrition elements nitrogen, phosphorus and potassium were introduced in a manure restriction module, with nitrogen as a fixed restriction that balances the production of nitrogen by animals and the possible uptake by crops.

7 Results of the economic and ecological evaluation of soil conservation policies

7.1 Finding relevant options for policy analysis

A clear framework is important for the selection of appropriate policy options, for the possibilities of combinations are too complex for all options to be calculated. Following such a framework for scenario design reduces the number of options to be calculated and makes it possible to focus on options that follow more or less common sense considerations.

It is important at this point to think again of the main goal of this study. It is not the aim of this study to find the optimal level of pollution (amount of soil eroded) i.e. an optimal solution for a social decision problem is not what is searched for here. However, the model can be used as a guideline on how the optimal management of soil conservation should be, when soil conservation is an exogenous restriction in the linear programming model. Therefore, in order to provide the **optimal low-cost solution** for a comparison, a stepwise **optimization scenario** ("social planner"-scenario) that generates results under optimal control conditions is added i.e. it is assumed that the maximum amount of soil erosion could be limited to a certain safe minimum standard directly through the decisions of the land users for the whole region.

As a main result, estimations for the costs of different soil conservation scenarios and the effectiveness of certain options should be available for decision makers. The aim is to find the total farm and budget costs of different approaches in soil conservation, mainly on the basis of the level of incentives that should be given (or not be given at all) and the measures that are supported. The efficiency criterion for the analysis of policy options is, as derived in chapter 4.5.2, a cost-effectiveness ratio, i.e. the costs of an option as a loss or gain in the gross-margin of the regional farm in the sample area. Within this approach, both budgetary and on-farm effects were analysed.

In the following, the specific options are outlined. The main variables that influence the set of options are:

- on-farm measures,
- policy instruments,
- targeting policies on the eligible area and
- transaction costs.

The main question is how soil erosion can be avoided or reduced in an economically efficient way with the help of **on-farm measures** such as reduced tillage systems or other soil conservation measures (see Chapter 4.7.3).

Policy instruments to enforce these measures are e.g. incentives or legal regulations. The reasons for focussing on these two options are outlined in Chapter 4.7.2. One of the most relevant variables on the instrument level, given their importance in the EU legislation are either incentives or legal restrictions for specific activities (Huylenbroeck and Whitby 1999). Subsidies for reduced tillage practices are offered in the model for all crops on all soil types. This option is often applied in the European Union agri-environmental programmes, so it reflects a common strategy for limiting soil erosion processes (Hartmann et al. 2006).

Furthermore, the variable share of the eligible land area for certain specific programmes (see Figure 14), is an issue that allows for more targeted soil conservation programmes compared to non-targeted conservation programmes, which are widely used in Germany today. The scenarios should comprise a comparison of a non-targeted scheme and a targeted scheme with focus on highly erosive field types, so that both methods may be analysed.

Therefore, on the policy level the analysed options are:

1. an incentive option,
2. a regulation option,
3. a targeted option.

For a comparison of different policy options, a similar level of erosion risk reduction is needed. In order to attain an obvious reaction of the model, different levels of reduction were tested under different policy conditions (see also Chapter 7.2). As a result of this process, a range of reduction of between 30 to 40 percent of the actual situation should be achieved for all policy options. This is assumed to be the **safe minimum standard** that society has agreed on and serves as a goal for policy comparison. A reduction range was chosen because the outcome of policies cannot be targeted to one specific number without changing the behaviour of the model. In the second step, the **threshold** suggestions from chapter 5.4.1 were used to benchmark the specific soil erosion risks of each crop that was found in each model solution. Thresholds (in terms of tons per hectare) for specific practices were applied in the regulation option that would ban high erosive crops from highly erodible field types.

The following definitions should clarify this:

Safe minimum standard	= general goal to be achieved through different policy options
Threshold	= benchmark for the result of a specific policy option

For the incentive options, an appropriate amount for the subsidy has to be calculated, while for the regulation options the choice of crops and eligible areas is the most important. The approach will be further described in the relevant chapters.

The degree to which **targeting policies for the eligible area** influences the efficiency of such a policy is the third issue. Chapter 5.4 describes options of how to define eligible areas and discusses the affect of different threshold on the size of an area that is eligible for soil conservation programmes.

From the soil category options, one approach was selected as the underlying soil database (Chapter 5.1.3.3, p.63). The soil data used was based on a 100 meter grid using the average for labelling and calculating the aggregated grid cell (100_meme).

Finally, on a separate level **the influence of transaction costs** is discussed (see Chapter 8).

Most of the analysis is based on a with-without comparison, where a status quo scenario without specific soil conservation measures is compared to scenarios containing a variation of some variables.

7.2 Brief description of the analysed policy options

First, a status quo solution (AG2000) based on the policy conditions of Agenda 2000 was calculated to generate a starting point solution of the model, as a way to check the model's plausibility against the original data. Based on these results a new scenario (CAP2013) was designed that contained the main policy changes that were introduced in the 2003 CAP-Reform (see p. 88). For this scenario, the conditions of the year 2013 were applied in order to avoid the implementation of the transitionally paid farm specific payments. A short outline of all scenarios is given in Table 27.

Then, a scenario that approaches the issue from a social planner viewpoint with complete information on soil erosion rates for all cropping practices was analysed. This "social planner scenario" allows one to limit the maximum amount of soil erosion in a stepwise manner to a certain threshold for the whole region ("social plannner"-scenario). This procedure provides information about the least cost solution for a soil conservation policy, including the generation of shadow prices of certain levels of soil erosion abatement. This "external knowledge" approach provides the "optimal" abatement price of each erosion level, which can serve as a benchmark for second best policies.

The other three policy options are based on the assumption of differing property rights with regards to the right to cause soil degradation through soil erosion.

Table 27: Description and abbreviations used for the analysed scenarios

Key	Scenario type	Remarks
AG2000	Status quo	Basic scenario for validation of model against empirical data; Policy framework based on Agenda 2000 settings
CAP2013	Basic scenario	Basic scenario without specific soil conservation policies based on terms for the year 2013
Social planner	Optimization	Erosion level is lowered stepwise - knowledge of total information on erosion rates of cropping practices assumed (social planner view)
Inc	Incentives untargeted	Incentives are given out to farmers for soil conserving practices
Inctar	Incentives targeted	Incentives are given out only on highly erodible field types (category 4, 5, 6)
Crop restrict	Crop restrictions targeted	Crops with a high erosion rate are restricted on highly erodible field types (category 5, 6)

Assuming that the right to degrade soil to a certain extent belongs to the land user, then an incentive option ("inc"-scenario) would be the more realistic option. In this scenario, an incentive is given for specific soil conserving cropping activities (i.e. reduced tillage); an instrument that is already being used in some federal states in Germany (Hartmann et al. 2006). In the terms of a Pigouvian approach, these incentives support practices with fewer externalities. The level of the incentive is also important for the adoption of a voluntary measure. It has to be high enough for a sufficient area to be enrolled in such an incentive scheme. In terms of efficiency (budget) it should be low enough to not result in too many windfall gains among the participants.

Then, the effects of a targeted incentive scheme were analysed ("inctar"-scenario) with the assumption that discrimination among land users was accepted. For this policy option, the incentives described above were targeted only towards field types with a high erosion risk instead of all field types. This option is considered more efficient since it concentrates incentives on erosive sites and reduces the risk of windfall gains on less erodible field types.

The level of both incentive options is calculated on the basis of a sensitivity analysis.

The last option is based on a similar set of property rights, but with the limitation that damages on soils through soil degradation can be prohibited by society, i.e. society claims the rights on the long-term fertility of soils and can therefore assign a specific use of the resource. This option is described by a restriction on crops ("crop restrict"-scenario) with a high erosion rate (higher than 2.5 t/ha/year, see Figure 12 p.69), which are banned from highly erodible field types. Row crops such as sunflowers, corn, potatoes and sugar beets are not allowed to be grown on these field types, regardless of the tillage technique used. This is based on the observation that such crops still have high erosion values even when a soil conservation practice is used (see Figure 13, p.70). This option could be combined with a compensation payment, but in this example only the command-and-control aspect of the policy was analysed.

7.3 Indicators

Table 28 gives an overview of the indicators that were analysed in the scenario calculations. The standard indicators for each analysed scenario are the total gross margin of the model region, the crop shares, changes in cropping practices, spatial changes where crops are grown, total erosion within the region, the ratio of changes in gross margin (i.e. costs of erosion control) and total erosion of the region as a cost-effectiveness indicator.

The average site specific erosion values and the individual crop erosion risk values show through the thresholds from chapter 5.4.1 whether the cropping practices used on such field types are still maintaining the long-term sustainability of soil use. By using the erosion risk values, one can check whether the crops do or do not exceed the weakest suggested threshold of 8 tons/ha/year of soil erosion risk. Each policy option is evaluated for whether it achieves this criteria.

Table 28: Analysed indicators in each scenario

Indicator	Remarks
Erosion level	Total amount of erosion within the region (sum of estimated erosion rates of all grown crops)
Erosion reduction	Reduction as compared to the CAP-scenario
Site specific erosion rates	Erosion rates per crop and site
Gross margin incl. Subsidies (GM)	Total gross margin of the model region as the aggregated result of all activities (max of objective function)
Area under conservation scheme	Area receiving payments or under regulation
Budget costs (BC)	Payments to farmers for soil conservation programmes
Net Gross Margin (GM-BC)	Total gross margin of the model region minus budget costs
On-farm costs (Net GM change)	Net changes in gross margin after policy change without payments
Total costs (BC+GM Change)	The social costs of a policy as the sum of budget costs and on-farm costs
Total cost-effectiveness (reduced erosion)	Ratio of changes in gross margin (i.e. costs of erosion control) and total erosion of the region
Cost effectiveness based on budget costs	Effectiveness related only to the budget costs
Cost effectiveness based on-farm costs	Effectiveness related only to the farm costs
Crop shares	Total crop shares for the whole region
Changes in cropping practices	Even if crop shares do not change, changes are still possible in the way crops are cultivated (plough, no tillage, reduced tillage)
Spatial changes where crops are grown	In order to meet soil erosion reduction goals, the model can place activities on less erodible areas

Furthermore, the shadow price of the erosion restriction, generated in a parameterized erosion level scenario, can be used to calculate the regional farm's marginal abatement costs at each specific level of erosion.

Budgetary and on-farm costs are differentiated in order to show the costs that arise from a policy option. Therefore, in scenarios that implement incentive payments for soil conserving practices, these payments are subtracted from the region's total results for the calculation of the net gross margin. The difference between the status quo gross margin and the net gross margin after policy adoption describes the on-farm costs of a policy.

The sum of the budget costs and the on-farm costs of a policy makes up the total costs of a conservation policy. The budget costs are borne by society, while the on-farm costs are opportunity costs borne by the farmers that are to be compensated by the subsidy payments. Without the subsidy payments farmers would produce on a higher level of productivity.

For each scenario, the results concerning this framework will be given.

7.4 Status quo scenario: Agenda 2000

In the following, the modelling results based on the Agenda 2000 policy conditions are described und discussed. The crop shares of the model results of the region should match the data available for the region. However, differences can partly arise from the fact that the model does not take into account factors such as risk reduction through the planting of a wider variety of crops. Other deviations of the model from real data can be explained by unknown costs asserted to production (Howitt 1995) or differences in on-farm transaction costs that make farmers choose other options.

Table 29: Modelled and actual (2001) crop shares in percent on arable land

Crop group	Crops	Actual values of the district (Uckermark)		Modelled values in test region	
		Share by crops (%)	Grouped (%)	Share by crops (%)	Grouped (%)
	Winter wheat	32.59		22.55	
	Winter rye	6.08		19.44	
	Winter barley	10.61		3.09	
	Triticale	11.02		6.32	
	Oats	1.06			
Total cereals			61.36		51.41
Corn	Corn silage	6.58	6.58	3.67	3.67
Rape seed incl. non food crops	Rape seed	15.69		7.97	
	Set aside with non-food crops	4.37			
			20.06		7.97
Peas, lupines	Peas	0.45			
	Lupines	0.04			
Legumes			0.49		
Sugar beets	Sugar beets	3.97	3.97	3.98	3.98
Potatoes/others	Potatoes	0.01		0.12	0.11
	Others	3.08	3.09		
Set aside		4.45	4.45	32.86	32.86
Total			100.00	100.00	100.00

Source: Matzdorf et al. 2003, based on aggregated project data and own calculations

The comparison of the actual and modelled crop shares of the region shows certain differences (see Table 29). The shares of cereals differed both when compared as individual crops and as a group.

More rapeseed was grown in the actual situation while set aside was used to a much higher extent in the model than in reality. The first explanation for the difference is the divergence of the model region from the source region, for the real data is based on a much bigger administrative district. Even though the model region was selected as a smaller sample of the total area of the district, some deviation can occur in terms of soil quality and other natural site conditions. This has effects on the preferability of some crops for certain parts of the region.

Another reason for the difference is the economic characteristic of LP-models that has led to the extended use of the set aside option; most of the poorer soils in the modelled result are not used as cropland. In reality, other non-economic reasons are preventing farmers from using this option at the optimal level even though it would economically be the better alternative. Since the LP model chooses generally for the set aside activity, the estimated overall erosion will be rather low. When less erosive crops such as cereals, rape seed and set aside are grouped together, both sides of the table show more or less the same share (> 80 % of arable crops). For potatoes and sugar beets, restrictions were introduced in the model with the assumption that restrictions are in the actual situation determined by quotas and contracts with processing factories.

Given these results, the starting point in terms of expected erosion amounts had to be analysed to ensure that the initial erosion rates in the modelled scenario are plausible. Therefore, the erosion rates of real and modelled crop shares were compared.

Table 30 shows the comparison of the expected average soil erosion risk per hectare of the highest soil erosion category (6) based on average crop shares found in the real and modelled data sets. This is a very rough approximation since the spatial location of the real crops is not known. However, for illustration needs this calculation is sufficient. The shares of the crop types were assigned with the soil erosion rate for a standard cropping practice. The resulting values for the average soil erosion risk show the erosion reducing effect of the set-aside option in the modelled result (1.23 t/ha/a as compared to 1.68 t/ha/a in the real crop distribution). The lower share of cereals, corn and rape seed in the modelled scenario caused less overall erosion in this soil category compared to the actual data. Note that this calculation is only a rough approximation of the underlying effects.

Both average soil erosion rates were above the strictest threshold of 1 t/ha/a. The soil erosion rates for corn exceeded a 4 t/ha/a threshold partially, which would be too high on a "25" soil quality type if the flexible threshold, which depends on soil quality is used.

Chapter 7 – Results of the economic and ecological evaluation of soil conservation policies

Table 30: Average erosion risk based on actual and modelled data on highly erodible field type "6"

Crop types	Actual share %	Modelled share %	Soil erosion rate t/ha/a	Rate*actual share	Rate*modelled share
Total cereals*	61.36	51.41	1.55 [1]	95.11	79.69
Corn	6.58	3.67	4.25	27.97	15.60
Rape seed	20.06	7.97	1.35	27.08	10.76
Legumes	0.49	0.00	0.69	0.34	0.00
Sugar beets	3.97	3.98	3.60	14.29	14.33
Potatoes	0.01	0.11	2.60	0.03	0.29
Others	3.08		1.00 [2]	3.08	0.00
set aside	4.45	32.86	0.07	0.31	2.30
Average soil erosion t/ha/a				1.68	1.23

[1] Winter wheat
[2] assumed average

Source: own calculations

Overall, the results of the model are satisfactory and sufficient as a starting point for further scenarios despite some disparity between model and reality.

The basic economic and soil erosion results of the Agenda 2000 scenario including the yields from livestock activities is €386 per hectare (total area of region) with a total erosion amount of 2680 tons and an average soil erosion rate of 0.18 tons/ha/a (see also Table 31). Conventional and reduced tillage were equally distributed in this scenario.

Table 31: Indicator values for the Agenda 2000 scenario

Average gross margin (Euro/ha)	386
Total erosion in region (t)	2,680
Average amount of soil erosion within the region (t/ha)	0.18
Conventional tillage (ha)	7,008
Reduced tillage (ha)	7,112
No undersown crops (ha)	9,423
Undersown crops (ha)	4,640
Intercrops (ha)	56

Source: own calculations

Note that these numbers are based on the overall region that includes pasture with very low soil erosion and therefore do only reflect the rates fairly for highly erodible field types. However, the comparison of this value to other scenarios serves as a benchmark for other soil conservation options.

Furthermore, the region has generally low average erosion levels but shows high values in some spots. This characteristic is illustrated in Table 32: When areas with the highest erosion risk and the highest quality (50_6) were analysed, an average soil erosion risk of 1.18 tons/ha/a would be found if all crops had the same share on this field type (1.23 t/ha/a for the modelled crop shares).

Nevertheless, sugar beets showed a maximum value of 15.08 tons/ha/a in some spots of the erosion risk assessment, which is above any of the suggested thresholds in Chapter 5.4.1. The maximum values were taken from the original database where values were calculated for each single aggregated grid cell (see chapter 5.2). Less erodible areas level out hot spots even in the highest erosion class. Therefore, the maximum values found in this class should be given some considerations.

Table 32: Average and maximum soil erosion risks for crops on soil category 50_6 in the Agenda 2000 scenario (highest erosion risk, best soil quality)

Crop	Average soil erosion rate of crop (t/ha/a)	Maximum soil erosion found in original database (t/ha/a)	Tillage type
Rapeseed	0.72	3.47	Reduced
Winter rye	0.68	3.29	Conventional
Winter wheat	0.89	4.25	Reduced
Sugar beets	3.61	15.08	Conventional
Average	1.18		

Source: own calculations

7.5 Basic scenario CAP2013 (CAP reform with decoupled payments)

The basis scenario serves as a reference scenario for the soil conservation policy options. In terms of property rights for soils, soil degradation is only restricted to good technical practice. Therefore, all common agricultural practices can be used by farmers. The scenario can also show, what influence a policy change from the Agenda 2000 conditions to decoupled payments under the CAP reform can have. Table 33 shows the total gross margin of the basic CAP2013 scenario is 445 €/ha, the total erosion amount is (4107 t) and the average amount of soil erosion within the region is 0.29 t/ha/a.

Table 33: Indicator values for CAP2013 scenario

Average gross margin (Euro/ha)	445
Total erosion in region (t)	4,107
Average amount of soil erosion within the region (t/ha)	0.29
Conventional tillage (ha)	10,757
Reduced tillage (ha)	3,362
Undersown crops (ha)	5,384
Intercrops (ha)	0

Source: own calculations

On high erodible soils (50_6) an average of 1.6 t/ha/a was estimated by the model (see Table 34), with sugar beets at 3.61 t/ha/a and sunflowers at 2.75 t/ha/a for average soil erosion risk rates. Both sugar beets (15.08 t/ha/a) and sunflowers (11.67 t/ha/a) showed high maximum erosion risk rates on specific grid cells in the region.

Table 34: Average and maximum soil erosion risks for crops on soil category with highest erosion risk and best soil quality in the CAP2013 scenario

Crop	Average soil erosion rate of crop (t/ha/a)	Maximum soil erosion found (t/ha/a)	Tillage type
Winter barley	0.57	2.81	Conventional
Rapeseed	0.72	3.47	Reduced
Winter wheat	1.48	7.73	Conventional
Sunflowers	2.75	11.67	Conventional
Sugar beets	3.61	15.08	Conventional
Weighted average	1.60		

Source: own calculations

The comparison of the Agenda 2000 scenario and the decoupled CAP2013 scenario showed a higher average gross margin for the CAP2013 scenario (see also Figure 20). The reason for the greater gross margin lies in the higher direct payments for pasture land, which overcompensate the loss of the direct livestock payments. Furthermore, even the formerly unused pasture land can now be used under the minimal care option (mulching). In the Agenda 2000 scenario some pasture land was not used due to the low profitability of this land type within the modelled region.

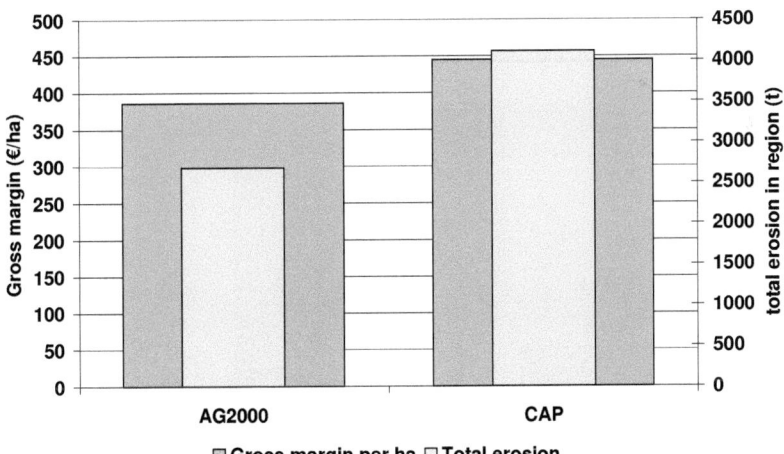

Source: own calculations

Figure 20: Gross margin (€/ha) and total erosion (tons) in model region - comparison of the Agenda 2000 and the CAP2013 scenario with decoupled payments

The increase in total erosion in the region of almost 55 percent can be explained by the higher share of row crops in the CAP2013 scenario (see Table 35). Note that this table includes the share of pasture land, which was not included in Table 29; the crop shares are therefore slightly different.

Chapter 7 – Results of the economic and ecological evaluation of soil conservation policies

The share of set-aside was even increased under the conditions of the CAP2013 scenario. Oilseeds such as rapeseed and sunflowers also had increased shares.

Table 35: Comparison of crop shares between Agenda 2000 and CAP2013 in percent of total area

Crop	Scenario	
	AG 2000 %	CAP2013 %
Corn silage	3.1	1.9
Hay 2 cuts	0.3	0.3
Non used pasture	4.7	-
Potatoes	0.1	0.1
Set aside	28.2	32.7
Set aside pasture	-	4.0
Silage 2 cuts	9.2	9.9
Sugar beets	3.4	3.1
Sunflowers	-	20.1
Triticale	5.4	-
Winter barley	2.7	1.6
Rapeseed	6.8	18.6
Winter rye	16.7	0.6
Winter wheat	19.3	7.2
sum	100	100

Source: own calculations

Figure 21 demonstrates that sunflowers, potatoes, corn silage and sugar beets sum up to more than 25 percent of the total crops of the region, with sunflowers contributing the most to this change.

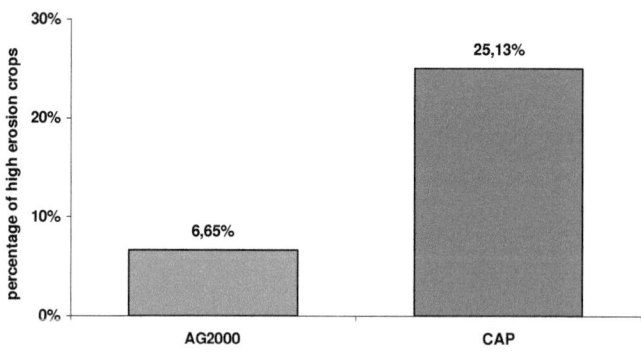

Source: own calculations

Figure 21: Share of high erosion crops (sunflowers, potatoes, corn silage and sugar beets) in the Agenda 2000 and the CAP2013 reform scenario

Furthermore, the share of reduced tillage practices was decreased in the CAP2013 scenario compared to Agenda 2000 from 10754 to 3362 hectares because the higher profit crops were less suited for reduced tillage (e.g. sunflowers).

Livestock production was also affected by the policy change (see Table 36). While the number of milking cows was reduced, more heifers were produced in the CAP2013 scenario. Bull fattening was no longer profitable without the per head subsidies of the Agenda 2000 support scheme. Pork production was reduced to 15,767 units in comparison to more than 20,000 (Agenda 2000).

Table 36: Livestock production in Agenda 2000 and CAP2013 scenario

Animal type	Agenda 2000	CAP2013
Milking cows	1,995	1,974
Heifers	1,537	2,190
Bulls, 1 to 1.5 years	914	
Fattening bulls, 1.5 to 2 years	914	
Hogs	20,223	15,946

Source: own calculations

As a result, the comparison of the Agenda 2000 and the decoupled CAP2013 reform scenario in the model region showed that an increase in soil erosion is possible. Given the risk assessment results for the other crops, even reduced tillage options for sunflowers would be unlikely to reduce the erosion risk to the levels found for rapeseed or cereals(see Figure 13, p.70).

The erosion rates for high erodible soil types are partly over a 1 ton/ha/a threshold but higher rates can be expected for specific spots (see Table 34, p.107).

7.6 Social planner scenario

7.6.1 Scenario description and trade-off curve

For this scenario, it is assumed that a social planner equipped with full information (and with no transaction costs involved) could change the level of erosion through optimal control techniques. The erosion rate is then a variable within the model that is directly influenced by the choice of crops and cropping practices assuming that complete information regarding the relations between cropping practices and the erosion output is known. Therefore, the model will most likely find the most cost-efficient solution for reducing the soil erosion risk.

This scenario is achieved through a stepwise limitation of the model's erosion restriction in parameterized runs. The total amount of erosion in the region was reduced within 20 steps to a level where the sum of estimated erosion caused by the cropping practices in the region was zero. The LP-model is programmed to find economically optimal solutions that meet the targeted lower erosion level for the region. As a result of this, a trade-off curve can be drawn showing the relationship between the economic performance of the region and the related soil erosion rates. Figure 22 shows that for a wide erosion range the reduction of the gross margin is very limited and only decreases at drastically lowered erosion levels. The underlying adjustments within the model that caused the shape of this trade-off curve will be analysed in the following.

Chapter 7 – Results of the economic and ecological evaluation of soil conservation policies

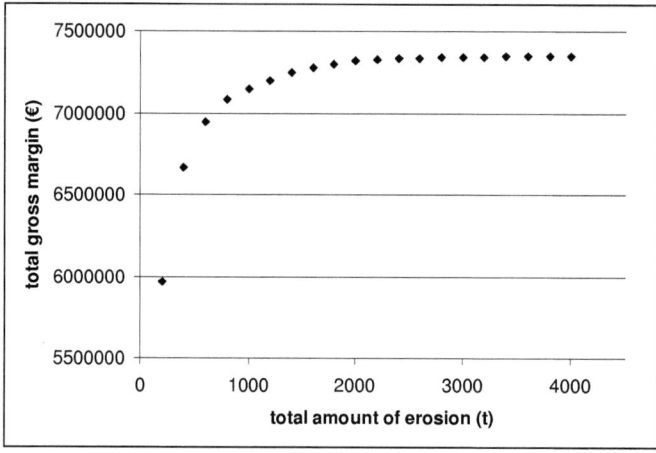

Source: own calculations

Figure 22: Trade-off curve between total erosion levels and total gross margin in region based on a parameterized model run with increasing limitation on the erosion level

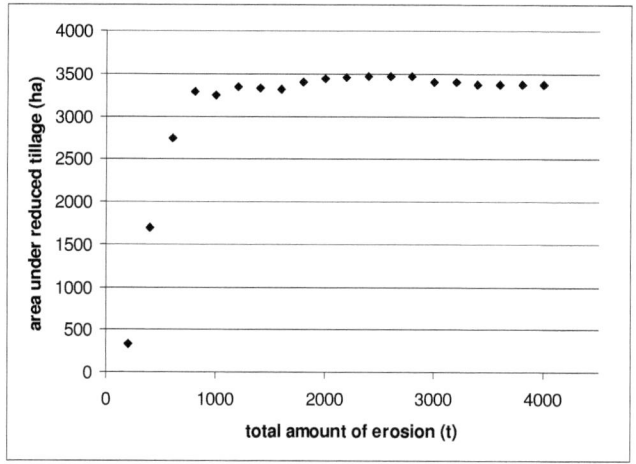

Source: own calculations

Figure 23: Share of reduced tillage with increasing levels of soil conservation in the model region (social planner scenario)

The corresponding share of reduced tillage decreased paradoxically with increasing erosion reduction levels (Figure 23); one of the reasons for the huge decrease of reduced tillage is that set-aside in this model was established using plough tillage for phyto-sanitary reasons. However, set-aside has one of the lowest erosion risk potentials and was therefore used mainly to achieve the lower erosion values.

Set aside increased continuously throughout the scenario runs from 30 to almost 70 percent at the lowest erosion level (Figure 24).

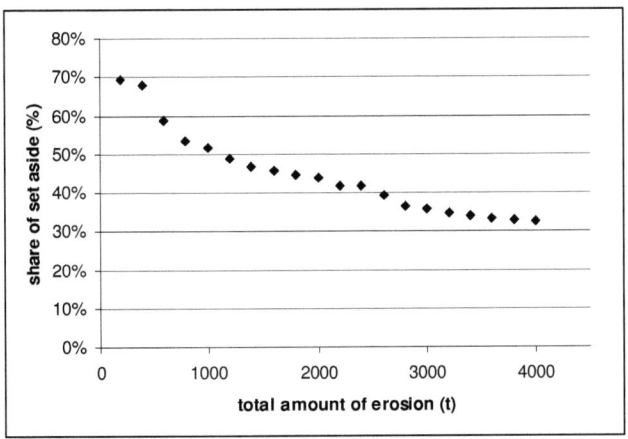

Source: own calculations

Figure 24: Share of set aside with increasing levels of soil conservation in the model region (social planner scenario)

In the basic scenario, only the soils with the lowest fertility were used for set-aside, as it would be the most profitable land use for this soil type. However, with higher levels of erosion control, set-aside became a measure for meeting this goal (even on better soils).

Table 37: Crop shares within the study region for selected levels of erosion in the social planner scenario

Crop		Total amount of erosion (tons)					
		200	1000	2000	2600	3000	4000
Set aside	%	69	52	44	39	36	33
Grassland silage	%	14	11	10	10	10	10
Winter rye	%	7	6	5	2	1	1
Rapeseed	%	1	18	19	19	19	19
Corn silage	%	1	2	2	2	2	2
Mulching Grassland	%	0	3	4	4	4	4
Potatoes	%	0	0	0	0	0	0
Winter wheat	%	0	4	4	7	7	7
Hay 2 cuts	%	0	0	0	0	0	0
Sunflowers	%	0	1	7	12	17	20
Winter barley	%	0	1	2	2	2	2
Sugar beets	%	0	2	3	3	3	3
Not used	%	7	0	0	0	0	0
Total	%	100	100	100	100	100	100

Source: own calculations

Table 37 gives a complete overview of the crops shares grown in the selected scenarios of erosion prevention. Rapeseed, sugar beets and sunflowers shares decrease, while winter rye share increases when the erosion risk is to be reduced. However, set aside is the only way for the model to achieve the highest levels of soil conservation. Generally, the erosion reduction effect of a crop with a naturally lower erosion rate is usually stronger than the effect of a high erosive crop cultivated with a soil conservation measure (see also Figure 13, p. 70). Therefore, the erosion reduction effect of crop selection overweighs the effects of most conservation practices.

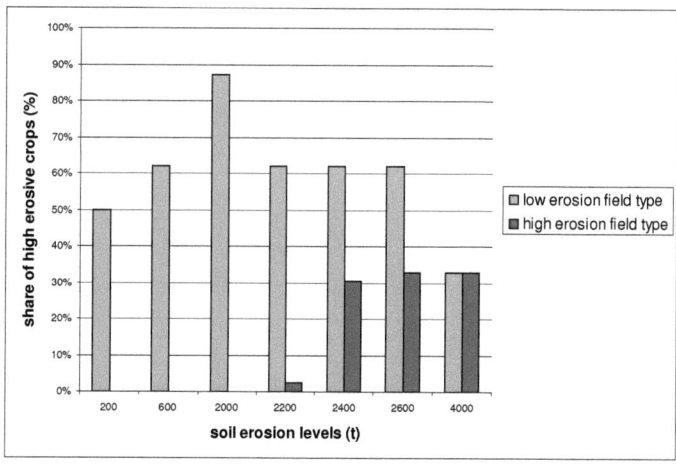

Source: own calculations

Figure 25: Share of high erosive crops on high and low erosion risk field types on selected levels of erosion in the social planner scenario

Figure 25 shows the shift of erosive crops from the most to the least erodible soil types when a higher soil conservation level is set through the erosion restriction. In the basic solution (4000 tons, on the left side of the x-axis) both soil types carry the same share of high erosive crops, such as sunflowers, potatoes, corn silage and sugar beets. When the restriction on soil erosion forces the model to adopt a slightly higher soil conservation level, high erosive crops from other soil categories shifted to the less erodible soil type, while the most erodible soil type kept its share of high erosive crops (2600 tons). At 2400 tons, the share of high erosive crops on the most erodible soil type had to be reduced in order to meet the even higher conservation level. Starting at 2000 tons, high erosive crops were not grown anymore on highly erodible soil types, while the less erodible soil type still kept a high share of 87 percent, a limit which was set by phyto-sanitary restrictions. At 600 tons, this share was reduced down to 62 percent, while at an erosion level of 200 tons only 50 percent of high erosive crops comprised mostly of corn silage were grown on the low erosive field type within the region. This example describes the general shift of high erosive crops to less erodible soil types during the parameterization runs.

In general, the behaviour of the model when faced with an exogenous reduction of erosion levels can be summarized as followed:

1. places high erosive crops in less erodible soil categories
2. replaces high erosive crops with less erosive ones
3. sets aside highly erosive field types

Reduced tillage plays only a minor role, when erosion has to be limited for the model region (see Figure 23, p.110). A certain share is generally adopted because of economic advantages (less labour needed).

The low impact on the economic situation of the model region towards even higher limitation of the overall erosion is mainly related to the flexibility of the model to shift crops between field types within the region. The re-organisation of crops at zero costs is only possible in a model. In reality this would cause coordination problems among different farms. Furthermore, given the general low soil qualities of this region, the gross margin for set aside is actually close to crops like winter wheat, so the exchange of crops on medium quality soil types does not cause drastic changes in the total gross margin.

7.6.2 Calculation of a shadow price for soil erosion

In LP-models, the calculated shadow price per unit of a scarce resource is the price a producer would pay for this resource if he had the option to buy more of this input. Figure 26 shows how the shadow price rises with higher levels of erosion reduction, which corresponds to the trade-off curve in the chapter before.

The reason for the non-linear increase of the shadow price is that highly erosive crops are only placed on less erodible field types, since this option is available at almost zero costs, abstracting from the fact that coordination efforts for the shifting of crops would occur in reality. If the region has to meet higher levels of erosion reduction, more expensive soil conservation measures would have to be applied. More profitable crops would only be banned from the region if erosion has to be absolutely avoided. Deriving an abatement cost function of soil erosion would be the basis for an economically driven choice of the optimal erosion level based on the theory of environmental economics. However, the direct adjustment of erosion rates is just not possible in real life. Therefore, in this example the shadow price serves only as a benchmark for other policy options, where the more efficient solution lays.

Chapter 7 – Results of the economic and ecological evaluation of soil conservation policies

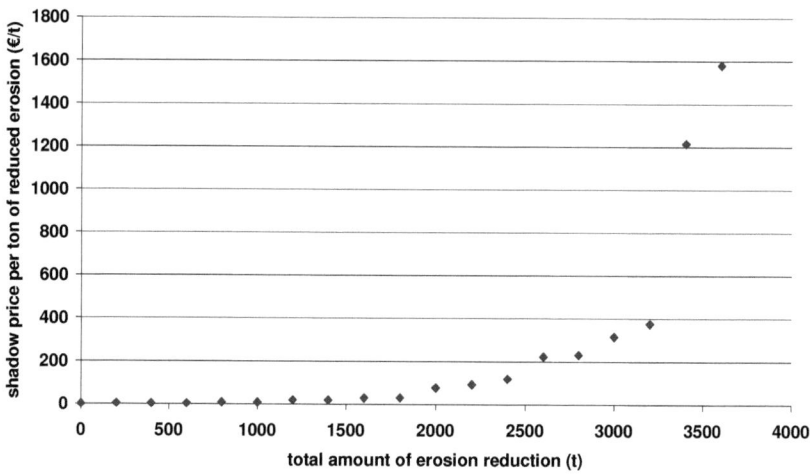

Source: own calculations

Figure 26: Marginal costs per ton of reduced erosion in the region; based on the shadow price per ton of reduced total erosion at each step of total erosion restriction

7.6.3 Results of the benchmark scenario

The results of the scenario run with an erosion level of 2600 tons in the region will be used for comparison with the policy scenarios. This level represents the range of 30 to 40 percent of the actual situation, which is stated as the goal of erosion reduction.

Table 38: Indicator values of the social planner scenario (2600 tons erosion level) compared to the basic scenario (CAP2013)

Indicator	Social planner	CAP2013
Average gross margin (Euro/ha)	444	445
Total erosion in region (t)	2,600	4,107
Average amount of soil erosion within the region (t/ha)	0.18	0.29
Conventional tillage (ha)	10,655	10,757
Reduced tillage (ha)	3,465	3,362
Undersown crops (ha)	6,450	5,384
Intercrops (ha)	0	0

Source: own calculations

Table 38 shows the indicators of this scenario in comparison to the results of the basic scenario. The comparison of gross margins shows that only a small change in the gross margin is needed to reach a much lower level of soil erosion risk. Reduced tillage and undersown crops increase slightly but the shifting of erosive crops to less erodible soils has a considerable effect (see Figure 25 p.112). Table 37 (p.111) illustrates the crop shares for the 2600t level in comparison to the 4000t scenario.

Main changes are in the increase of set aside and winter rye, and the reduction of sunflowers to achieve a lower erosion level.

7.7 Policy scenario results

In the following chapters, the basic results of the policy scenarios described before (comprised of an untargeted incentive scheme, a targeted scheme on more erodible soil types and a targeted scheme on the restriction of high erosive crop types) are described. After a brief description of each scenario result, a comparison of the scenarios is done, since most information is gained through the analysis of the changes that come with them. As mentioned before, all policy options were designed to improve the soil erosion situation found in the basic CAP2013-Reform scenario by at least 30 percent, which represents the assumed safe minimum standard.

7.7.1 Untargeted Incentive Scheme

7.7.1.1 Scenario description

In this scenario, a subsidy for the reduced tillage practice is offered in the model for all crops on all soil types. This option is often applied in the European Union agri-environmental programmes, so it reflects a common strategy for limiting soil erosion processes (Hartmann et al. 2006). The level of the payment is derived through a sensitivity analysis of the level of the incentive payment, so that the amount of payment which reduces erosion to the targeted amount of at least 30 percent may be found (see Chapter 7.1 and Chapter 7.7.1.1). The resulting effects of this payment level are then described based on the aforementioned scenario indicators.

For each new step in the sensitivity analysis, a value slightly higher than the upper limit of the reduced tillage incentive payment is used as the new value for the next run. This procedure is repeated until at least a 30 percent reduction of the total soil erosion in the region is achieved.

Figure 27 shows that the reduced tillage payments do not influence the amount of erosion until the payment reaches a level over 60 €/ha. This amount more than doubles the area covered by the scheme from 3570 to 9224 ha. However, the corresponding amount of erosion only drops to the targeted amount when a payment level of 68 €/ha is chosen.

The result shows that a certain share of reduced tillage practices is already being used in the basic CAP2013-solution. Payments for the further expansion of this practice only force the model to adopt the reduced tillage practice for crops that do not improve the overall situation of soil conservation. Only payments over 68 €/ha cause the switch of another share of crops to reduced tillage. Therefore, this scenario describes how reduced tillage payments would affect a region, where reduced tillage is already a partly adopted cropping technique.

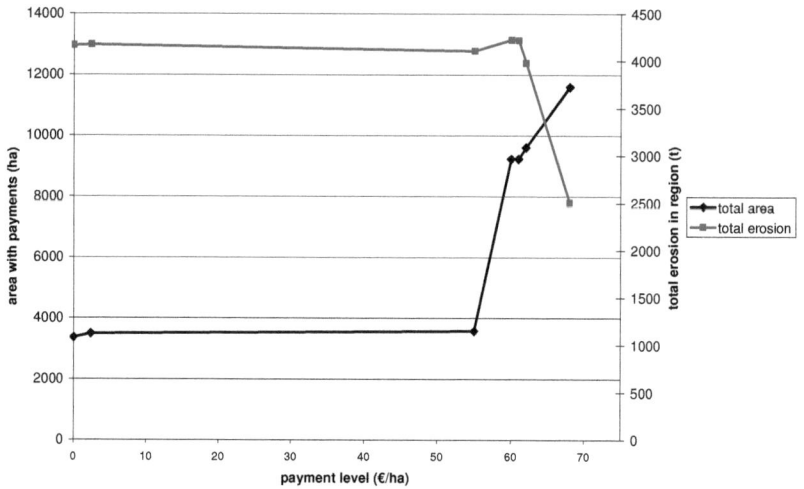

Source: own calculations

Figure 27: Effect of the payment level for reduced tillage on the area under the conservation scheme and the total erosion in the region; each payment level corresponds to the upper limit of the sensitivity analysis of the preceding calculation step

7.7.1.2 Basic results

As a result of the incentive option with the payment of 68 €/ha for reduced tillage practices, almost 12000 ha of the region got included under the reduced tillage practice and would therefore be enrolled in this conservation programme. The area covers 85 percent of the arable land in the region. In the basic result (CAP), only 3362 ha were cultivated with reduced tillage practices.

Table 39: Share of reduced tillage in the untargeted incentive scenario (68 €/ha for reduced tillage) compared to the CAP2013-scenario

Tillage type	CAP2013	Incentive untargeted
Reduced tillage (ha)	3,362	11,597

Source: own calculations

The total erosion in the region was reduced by almost 1600 tons. The gross margin rose to €7.6 M because of the high amount of subsidies received (€788,574). When the budget costs (payments received in the region) were subtracted from the total gross margin, a net gross margin of €6.8 M remained. This is about €0.5 M less than in the basic solution with standard CAP conditions (CAP2013). This difference is defined as the on-farm costs, since these costs describe the opportunity costs of the agricultural changes in the region due to the adoption of the policy.

The sum of on-farm costs and budget costs describes the total costs of a policy from society's viewpoint. These total costs are used with the erosion reduction in the region in the cost effectiveness ratio. As a result, €805 per one ton of avoided soil erosion were spent in this scenario.

Table 40: Indicator values of the untargeted incentive scenario (68 €/ha for reduced tillage) compared to the CAP2013-scenario

Scenario		CAP2013	Untargeted Incentives
Erosion level	t	4,107	2,510
Erosion reduction	t	0	1,597
Gross margin (incl. Subsidies) (GM)	€	7,330,440	7,621,731
Area under conservation scheme	ha	0	11,597
Budget costs (BC)	€	0	7,88,574
Net Gross margin (GM-BC)	€	7,330,440	6,833,157
On-farm costs (Net GM change)	€	0	497,283
Total costs (BC+GM Change)	€	0	1,285,857
Total cost-effectiveness (€/t reduced erosion)	€/t	0	805

Source: own calculations

Table 41 gives an overview of the changes in crops grown in the model region. Major changes were seen in the reduction of sunflowers, since the reduced tillage option for this crop is not possible both in reality and in the database. This crop was therefore replaced by other reduced tillage crops. Set aside had even more increase to the level of almost 44 percent. Small changes also occurred in the production of cereals, where winter barley increased slightly, while winter rye and winter wheat held more or less the same share. Rapeseed and sugar beet shares also increased slightly in the region.

Table 41: Changes of crop shares in the untargeted incentive scenario (68 €/ha for reduced tillage) compared to the CAP2013 scenario

Crop		CAP2013	Untargeted Incentives
Set aside grassland	%	4.01	4.72
Hay, 2 cuts	%	0.34	0.34
Sunflowers	%	20.09	7.19
Potatoes	%	0.10	0.10
Corn, silage	%	1.85	2.35
Silage, 2 cuts	%	9.92	9.21
Rotational set-aside	%	32.69	43.89
Winter barley	%	1.63	2.08
Rapeseed	%	18.56	19.33
Winter rye	%	0.55	0.36
Winter wheat	%	7.19	7.19
Sugar beets	%	3.06	3.24

Source: own calculations

7.7.2 Targeted Incentive Scheme

7.7.2.1 Scenario description

This scenario shows the effect of a soil conservation programme that uses targeted incentives for the adoption of soil conservation measures in areas with an elevated erosion risk. In this scenario, the same subsidy (68 €/ha) as in the untargeted scheme was only available on field types of the groups 4, 5 and 6 (higher erosion classes), making more than 11,000 ha of the region eligible for the subsidy payment. A test run with only soil groups 5 and 6 was also done, but the eligible area combined with the potential of the conservation measure was not large enough to reach a comparable erosion reduction of 30 percent. Like before, the measure to be performed for this scheme comprises all crops grown with a reduced tillage system.

7.7.2.2 Basic results

The total erosion potential in the region was reduced by more than 38 percent compared to the CAP2013 scenario, while the total gross margin increased up to €7.57 M. In this scenario 9421 ha of the possible 11,000 ha were under the conservation scheme. The budget costs added up to €640,639. The net gross margin (Total gross margin minus budget costs) was also lower than in the CAP2013 scenario. This led to on-farm costs of €401,409. Due to the on-farm costs, the total costs of the targeted scheme were €1.04 M, which led to the cost-effectiveness of €557 per ton of reduced erosion.

Table 42: Indicator values for the targeted incentive scenario (68 €/ha for reduced tillage) compared to the CAP2013-scenario

Scenario		CAP2013	Targeted Incentives
Erosion level	t	4,107	2,531
Erosion reduction	t	0	1,576
Gross margin (incl. Subsidies) (GM)	€	7,330,440	7,569,671
Area under conservation scheme	ha	0	9,421
Budget costs (BC)	€	0	640,639
Net Gross margin (GM-BC)	€	7,330,440	6,929,032
On-farm costs (Net GM change)	€	0	401,409
Total costs (BC+GM Change)	€	0	1,042,048
Total cost-effectiveness (€/t reduced erosion)	€/t	-	661

Source: own calculations

The crop shares in the targeted incentive scenario developed as follows (Table 43). Sunflowers were reduced by more than 50 percent, while the share of set aside increased. Other crops showed only little changes towards this policy option.

Chapter 7 – Results of the economic and ecological evaluation of soil conservation policies

Table 43: Crop shares in the targeted incentive scenario (68 €/ha for reduced tillage) compared to the CAP2013-scenario

Crop		CAP2013	Targeted Incentives
Set aside grassland	%	4.01	4.72
Hay, 2 cuts	%	0.34	0.34
Sunflowers	%	20.09	9.73
Potatoes	%	0.10	0.10
Corn, silage	%	1.85	2.35
Silage, 2 cuts	%	9.92	9.21
Rotational set-aside	%	32.69	41.66
Winter barley	%	1.63	1.92
Rapeseed	%	18.56	19.19
Winter rye	%	0.55	0.39
Winter wheat	%	7.19	7.19
Sugar beets	%	3.06	3.21

Source: own calculations

The underlying shifts of the cropping practices will be discussed in the overall comparison of the policy options.

7.7.3 Targeted crop restrictions

7.7.3.1 Scenario description

This scenario analyses the effects of restriction on highly erosive crops on targeted high erodible sites. It follows the assumption that most reduced tillage practices do not reduce the risk of erosion as well as change in crop selection (see Figure 13). Therefore, this scenario should prove whether it is more preferable to restrict the choice of certain crops by law as opposed to voluntary approaches such as subsidies on reduced tillage.

In the scenario, row crops such as corn, sunflowers, potatoes and sugar beets were not allowed on the field types 5 and 6, which cover 3,578 ha of the region. The limitation on these field types was proven to be sufficient in meeting the goal of a minimum 30 percent reduction in soil erosion risk. No other policy restrictions were implemented.

7.7.3.2 Basic results

Table 44 shows the results of the row crop restriction scenario. With the application of this policy option the reduction of 1,505 tons was achieved as compared to 4,107 tons in the CAP2013-scenario within the region. The total gross margin of the region was reduced by €14,340. Budget costs in terms of payments to farmers did not occur in this scenario and administration costs were not accounted for in this analysis. However, assuming the compensation for the on-farm costs had to be decided in a political process, the amount of compensation does not affect the resulting land use, since the restriction of crops is a mandatory regulation. Therefore, the amount of compensation would be part of a negotiation process during the implementation process.

The on-farm costs are therefore equal to the reduction of total gross margin (€14,340). The cost-effectiveness in terms of reduced soil erosion is then 10 €/ton.

Table 44: Indicator values for the row crop restriction scenario with restricted cultivation of row crops on highly erodible field types compared to the CAP2013-scenario

Scenario		CAP2013	Row crop restrictions
Erosion level	t	4,107	2,602
Erosion reduction	t	0	1,505
Gross margin (incl. Subsidies) (GM)	€	7,330,440	7,316,100
Area under conservation scheme	ha	0	-
Budget costs (BC)	€	0	-
Net Gross margin (GM-BC)	€	7,330,440	7,316,100
On-farm costs (Net GM change)	€	0	14,340
Total costs (BC+GM Change)	€	0	14,340
Total cost-effectiveness (€/t reduced erosion)	€/t	-	10

Source: own calculations

Compared to the CAP2013-scenario, only sunflowers were reduced in a noticeable way (from 20 % to 15 %) (see Table 45).

Table 45: Crop shares for the row crop restriction scenario (restricted cultivation of row crops on highly erodible field types) compared to the CAP2013-scenario

Crop	CAP2013	Row crop restriction
Set aside grassland	3.91 %	3.91 %
Hay	0.34 %	0.34 %
Silage	10.02 %	10.02 %
Sugar beets	3.06 %	3.06 %
Sunflowers	20.09 %	15.44 %
Potatoes	0.10 %	0.10 %
Corn silage	1.87 %	1.87 %
Winter barley	1.58 %	1.96 %
Winter rye	0.58 %	2.02 %
Winter wheat	7.19 %	7.19 %
Rapeseed	18.56 %	17.98 %
Set aside	32.69 %	36.10 %

Source: own calculations

Other crops such as winter barley, winter rye and set aside took over the share of the reduced sunflower area. For all other crops, almost similar results were achieved in this scenario, since only a smaller share of the agricultural area was affected by the crop restriction. This allows for the spatial compensation of the restriction effects through the switching of crops between field types. The share of reduced tillage decreased slightly in this scenario to 3,281 ha compared to the CAP2013-scenario.

7.8 Discussion of the modelling results

7.8.1 Overall comparison of scenarios

This chapter compares the results of the analysed policy scenarios using the indicators that were derived in the chapters before. This comparison serves as the basis for the discussion of the results and as a starting point for the application of a new institutional economics analysis in the following chapters.

Table 46: Overview of indicator values of the different policy scenarios (CAP2013, Untargeted Incentives, Targeted Incentives, Row crop restrictions, Social planner)

		CAP2013	Untargeted Incentives	Targeted Incentives	Row crop restrictions	Social planner
Erosion level	t	4,107	2,510	2,531	2,602	2,600
Erosion reduction	t	0	1,597	1,576	1,505	1,507
Gross margin (incl. Subsidies) (GM)	€	7,330,440	7,621,731	7,569,671	7,316,100	7,319,301
Average GM/ha total area	€	445	415	421	444	444
Area under conservation scheme	ha	0	11,597	9,421	0	0
Budget costs (BC)	€	0	788,574	640,639	0	0
Net Gross margin(GM-BC)	€	7,330,440	6,833,157	6,929,032	7,316,100	7,319,301
On-farm costs (Net GM change)	€	0	497,283	401,409	14,340	11,139
Total costs (BC+GM Change)	€	0	1,285,857	1,042,048	1,4340	11,139
Total cost-effectiveness (reduced erosion)	€/t	0	805	661	10	7
Cost effectiveness based on budget costs	€/t	0	494	407	0	0
Cost effectiveness based on-farm costs	€/t	0	311	255	10	7

Source: own calculations

The results for the selected policy scenarios show that all three options achieved a similar reduction in soil erosion compared to the CAP2013 scenario, which was the target range for soil erosion reduction. Both incentive options lowered the erosion level of the region with only little advantages for the targeted version, while the targeted row crop restriction showed slightly higher but comparable results (see Figure 28 and Table 46). Note that the erosion level was not fixed in the model but was aimed at with several tests through the adjustment of the incentive level or the size of the area that is affected by the row crop restriction. Therefore, the levels varied slightly between 1,500 and 1,600 t in the region.

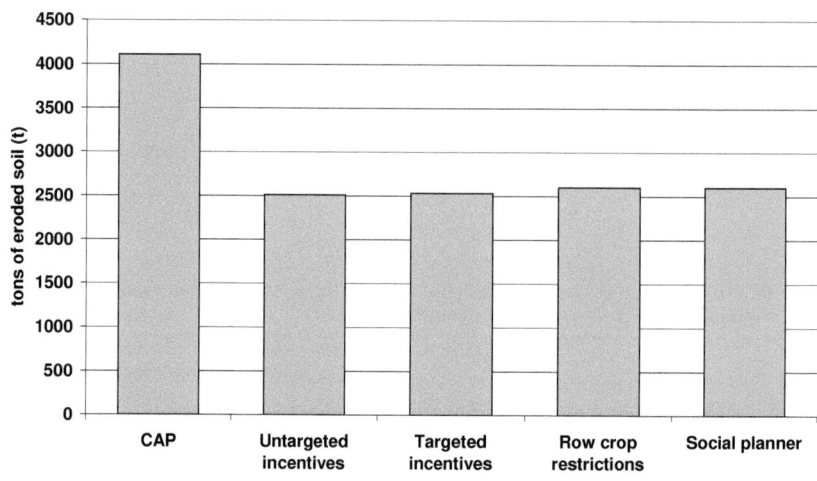

Source: own calculations

Figure 28: Total soil erosion in the model region under different policy options

Due to the incentives for reduced tillage practice in the "Incentive" scenarios, both scenarios of this kind showed unsurprisingly high increases in the share of reduced tillage for the region (see Table 47). The row crop restriction scenario achieved the erosion reduction only through the shift of erosive crops to less erodible field types, while the share of reduced tillage practices was even lower than in the CAP2013-scenario. The social planner scenario increased the reduced tillage slightly compared to CAP.

Table 47: Share of reduced tillage in the different scenarios

	CAP2013	Untargeted incentives	Targeted incentives	Row crop restrictions	Social planner
Reduced Tillage (ha)	3,362	11,597	9,421	3,281	3,465

Source: own calculations

Table 48 gives a summary of the crop shares grown in the different policy options. While the share of grassland was constant with only slight changes in its usage, the share of row crops, sun flowers in particular varied strongly. In the CAP2013 solution sunflowers reached a share of 20 percent but decreased to almost seven percent in the untargeted incentive scenario. Set aside increased in all conservation scenarios with a maximum share of more than 43 % in the untargeted incentive scheme. Other crops showed only small shifts in their shares.

Chapter 7 – Results of the economic and ecological evaluation of soil conservation policies

Table 48: Overview of crop shares under different policy options

		CAP2013	Untargeted incentives	Targeted incentives	Row crop restrictions	Social planner
Set aside grassland	%	4.01	4.72	4.72	3.91	3.91
Hay, 2 cuts	%	0.34	0.34	0.34	0.34	0.34
Sunflowers	%	20.09	7.19	9.73	15.44	12.22
Potatoes	%	0.10	0.10	0.10	0.10	0.10
Corn, silage	%	1.85	2.35	2.35	1.87	1.87
Silage, 2 cuts	%	9.92	9.21	9.21	10.02	10.02
Rotational set-aside	%	32.69	43.89	41.66	36.10	39.17
Winter barley	%	1.63	2.08	1.92	1.96	1.72
Rapeseed	%	18.56	19.33	19.19	17.98	19.19
Winter rye	%	0.55	0.36	0.39	2.02	1.56
Winter wheat	%	7.19	7.19	7.19	7.19	6.69
Sugar beets	%	3.06	3.24	3.21	3.06	3.21

Source: own calculations

To analyse the costs of the three policy options (see Table 46), a distinction was made between the on-farm costs that stem from losses on the farm due to changes away from an optimal solution (opportunity costs) without any policy intervention and the budgetary costs of a policy which comprise only the expenses of incentives paid to the farmers for reasons of simplicity. As described before, compensation payments might be possible in a legal restriction scenario but such compensations are not crucial for the outcome of the policy option (see Chapter 7.7.3.1).

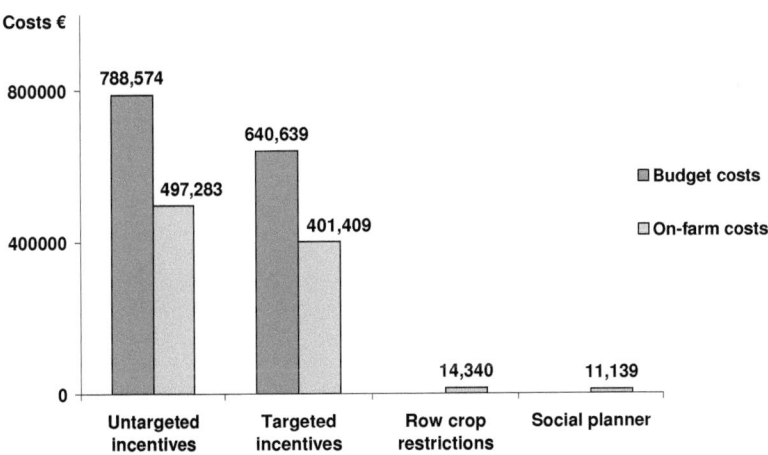

Source: own calculations

Figure 29: Costs of different policy options

Figure 29 illustrates the highest costs for an untargeted incentive option both for the on-farm costs as well as the budget costs (see also Table 46). The higher on-farm costs for the untargeted compared to the targeted option can be explained by the bigger leeway the targeted option has in the use of highly profitable crops for reducing on-farm costs. Since payment is available on all field types in the untargeted option, there is no need to consider whether a ploughing tillage system is more profitable. However, from a regional farmer's viewpoint, the policy with the highest total gross margin, which is the untargeted incentive scheme, is preferable. For the row crop restriction option, no direct budget costs arise under the given assumptions, so that only on-farm costs are realised through a change of production practices. This holds true as well for the social planner scenario, which caused almost similar on-farm costs as the row crop restriction scenario.

The cost-effectiveness of the different policy options can be based either on the total costs (the sum of on-farm costs and budget costs) or separately on the possible budget costs and on-farm costs for each option in relation to the physical output (erosion reduction) the policy option provides. Both alternatives (either as total costs or itemized costs) are either biased towards the incentive or the restriction option. However, the comparison of both cost types also provides valuable insights, for it shows where the costs occur. For now, both options serve as appropriate indicators for the effectiveness of a policy.

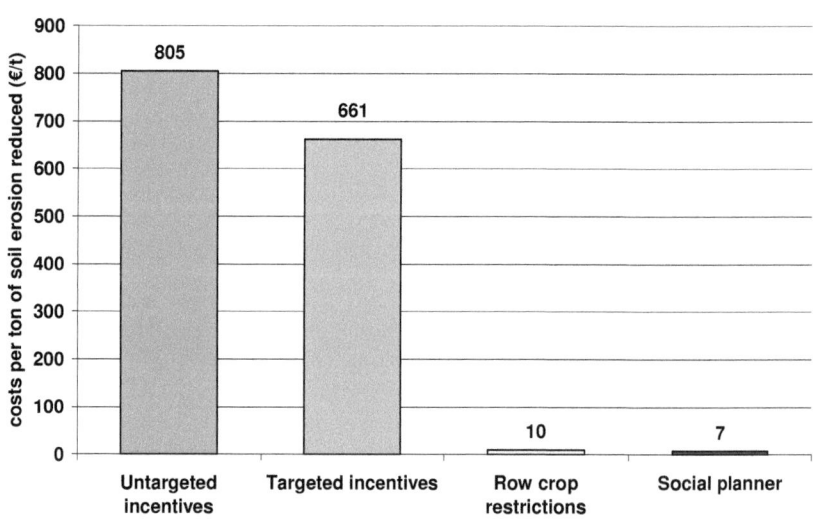

Source: own calculations

Figure 30: Cost-effectiveness of different policy options based on total costs

Using the total cost-effectiveness ratio for each policy, it is shown in Figure 30 that a targeted row crop restriction scenario is more efficient than both incentive options. It is remarkable how close the row crop restriction came to the cost-effectiveness of the social planner scenario that optimised land use while taking the total erosion risk into account.

Figure 31 shows the partial cost-effectiveness related to budget and on-farm cost. When only the budget costs were considered, the row crop restriction showed through definition "no" costs, while the targeted incentive scenario was proven the second best option. When the on-farm costs were compared, the row crop restriction option showed again the lowest distortion effects in terms of adaptation costs to a policy.

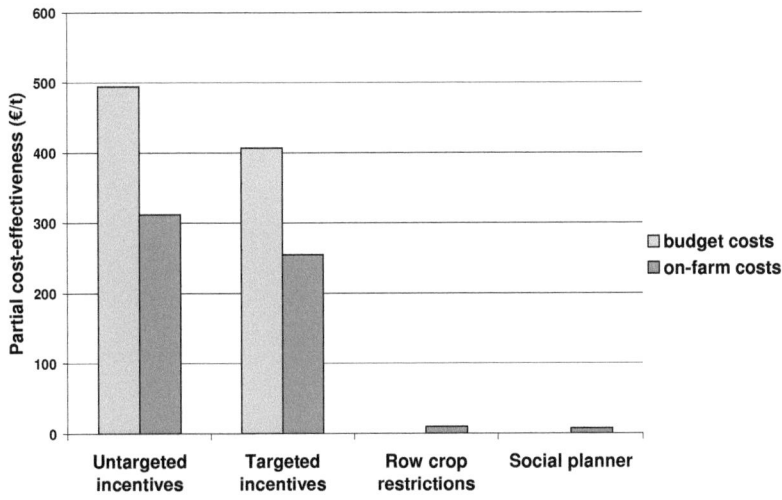

Source: own calculations

Figure 31: Partial cost-effectiveness of policy options based on budget costs and on-farm costs

7.8.2 Analysis of erosion levels on high erodible soil types

The applied bio-economic model provided detailed information on the crops grown on specific soil types. In the following paragraph the crop rotations selected by the model are described for the soil category with the highest soil quality type and the highest erosion risk (50_6) for the four analysed policy options and the social planner scenario.

Figure 32 shows the erosion risk values of the crop rotation of this category broken down to individual crops. The crop shares are a result of the optimization. The figure shows both the average value (columns) and the minimum and maximum value of each crop that was found in the soil erosion risk assessment database for the model region. Due to the different slope types in the soil categories, some grid cells showed erosion risk of up to 15 tons/ha/year.

Chapter 7 – Results of the economic and ecological evaluation of soil conservation policies

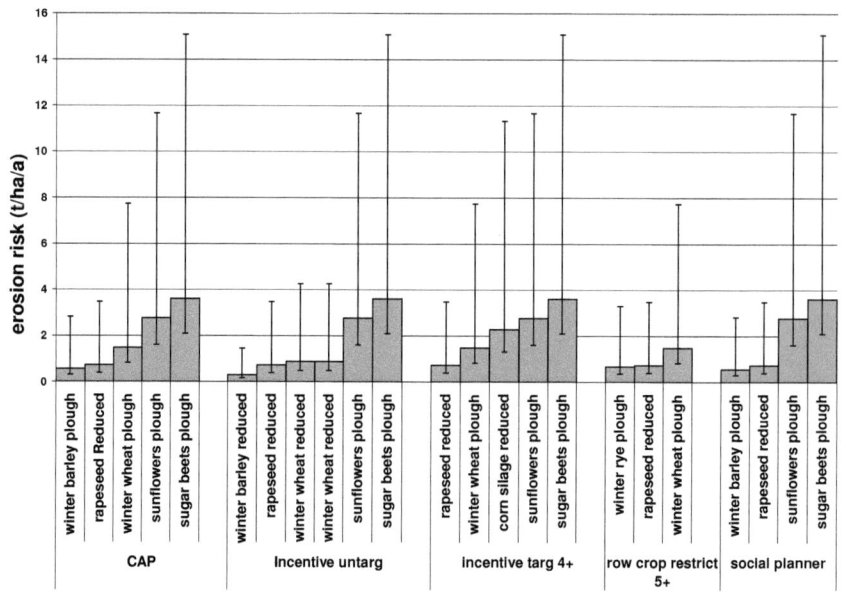

Source: Sattler 2007 and own calculations

Figure 32: Erosion risk of crops on high erodible field types for different scenarios shown as average (columns) and minimum and maximum values as found on single grid cells

For all policy options except for the row crop restriction, crops with a high erosion risk potential were also found on highly erodible soil types. High soil quality prevents the adoption of the reduced tillage options, which have in fact less variable costs but are less adopted for reasons of manure management within the model region[13]. Even though sugar beets were cultivated with the reduced tillage option, the average erosion risk was still at 2.0 t/ha with a maximum of 9.8 t/ha. This example shows that all the scenarios have a comparable erosion reduction over the region, but the extremes vary. High values of erosion risk can be found even in the social planner scenario for these soils. The reduction of soil erosion risk is achieved at this level of erosion prevention on less fertile soil categories.

A final analysis was done on the spatial positioning of crop types. Table 49 shows the crop shares on the highest soil erosion risk soil type with high soil quality under soil conservation policies in comparison to the CAP scenario (see also Figure 32). Sunflowers kept the same share, except for in the restriction scenario, where row crops are not allowed on this field type. Corn silage was only grown in the targeted incentive scenario, while the rapeseed share increased in all conservation

---

[13] Both sugar beets and corn silage are suitable for high manure uptakes. The reduced tillage options are designed for lower levels of manure application.

scenarios. Sugar beets were not reduced in the incentive scenarios but were not allowed in the restriction scenario. Rye served as the dominant crop in the restriction scenario. The share of winter wheat was not affected by the policies.

Table 49: Crop shares on highly erosive field type with good soil quality

Crops		CAP2013	Untargeted Incentives	Targeted Incentives	Row crop restrictions	Social planner
Sunflowers	%	25	25	25		25
Corn silage	%			17		
Rapeseed	%	25	25	25	25	25
Winter barley	%	17	17			42
Winter rye	%				50	
Winter wheat	%	25	25	25	25	
Sugar beets	%	8	8	8		8

Source: own calculations

7.8.3 Thresholds results

In order to find out which scenario meets the threshold suggestions of chapter 5.4.1, the modelling results were compared to the respective threshold values. The analysis was focussed on the soil erosion risk of the specific crops, to see whether values that surpass the demanded level of the threshold could be found.

Table 50: Soil erosion risks of crops in scenarios compared to the threshold values for soil erosion risk (++ = below threshold for maximum values, + = below threshold for average value, - = above threshold for average value)

Scenario	Tolerable soil erosion (TSE) <		
	1t/ha*a	Soil quality index / 8 (t/ha*a)	8t/ha*a
CAP2013	-	-	+
Untargeted incentives	-	-	+
Targeted incentives	-	+	+
Row crop restrictions	-	+	++
Social planner	-	+	++

Source: see Table 13 and own calculations

Table 50 lists the scenarios and the threshold values and indicates which scenario could meet a threshold. The threshold is not met, if one crop shows higher erosion rates than the benchmark of the threshold. For a more specific distinction, the maximum erosion risk found within the region data was also used for comparison. All scenarios were able to meet the 8t/ha/a threshold, since it was even meet by the CAP2013-scenario. No scenarios could meet the 1t/ha/a threshold, for at least one crop (mostly sugar beets) would have higher levels of erosion. Only the CAP and the untargeted incentive option could meet the soil quality related threshold.

When the maximum values found for erosion risk were used for comparison, even the 8t/ha/a threshold became a difficult level to reach. Only the row crop restriction and the social planner scenarios were able to keep erosion risk values below 8t/ha/a.

7.8.4 Spatial analysis of erosion rates under different policy options

The linkage of soil categories in the bio-economic model to the GIS-data allows the illustration of resulting erosion rates for each soil category in a map based on the average soil erosion risk for the region. For each category formed by the soil quality and erosion risk type, an average erosion value was calculated on the basis of the resulting crop rotation in this field type. Figure 33 shows the average erosion risk under the CAP2013 conditions. The map shows elevated erosion levels scattered over the whole region, reflecting the heterogeneity of soil erosion risk in the region, which was also shown in the initial map on soil erosion risk of the region (Figure 8, p.61). Note that the map in Figure 8 is based on an estimated average crop rotation and does not take into consideration, the fact that farmers may select the crop rotations according to the soil quality. Since the model calculated mostly set-aside for poor soil types, these soils showed a relatively low soil erosion risk even on high erodible sites. The highest average erosion rates were found on sites with high soil quality and high natural erosion risk due to slopes. Under the CAP2013 conditions these soils types were partly grown with sunflowers and sugar beets (see Table 49).

The untargeted incentive option (see Figure 34) lowered the erosion risk in the region. The highest erosion risk sites under the CAP2013 conditions improved under this option to less than 1.5 t/ha/year of erosion risk. The map shows an overall reduction in soil erosion risk for most sites as a result of the increase in reduced tillage practices.

Under the conditions of targeted incentive payments on reduced tillage the reduction of soil erosion risk is achieved in a different way (see Figure 35). Some hot spots still show erosion rates of the highest class, even though the total erosion amount was reduced to the same level as in the untargeted payment option. One reason for this is the cultivation of reduced tillage corn silage on the soil category with highest quality and highest erosion risk (see Table 49).

The restriction of row crops on high erosion soil categories resulted in the cut back of soil erosion risk on all sites (see Figure 36). No soil category exceeded the average soil erosion risk of 1 t/ha/year.

If a soil conservation policy that prevents any sites from still showing elevated soil erosion risk is needed, then row crop restriction would be more effective. However, if only an overall reduction within the region is aimed at, then all of the policies would be effective.

Chapter 7 – Results of the economic and ecological evaluation of soil conservation policies

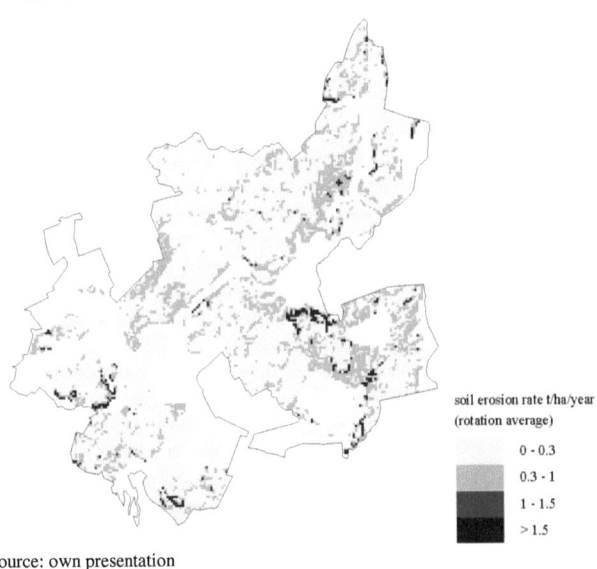

Source: own presentation

Figure 33: Average soil erosion risk for the region under the CAP2013 conditions

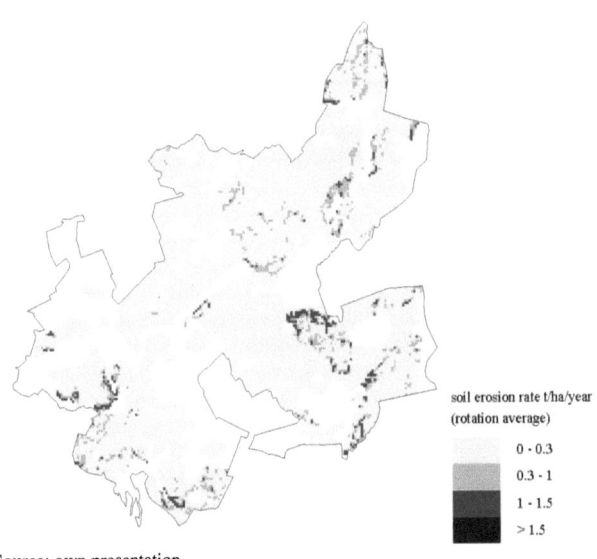

Source: own presentation

Figure 34: Average soil erosion risk for the region under untargeted incentive conditions

Chapter 7 – Results of the economic and ecological evaluation of soil conservation policies

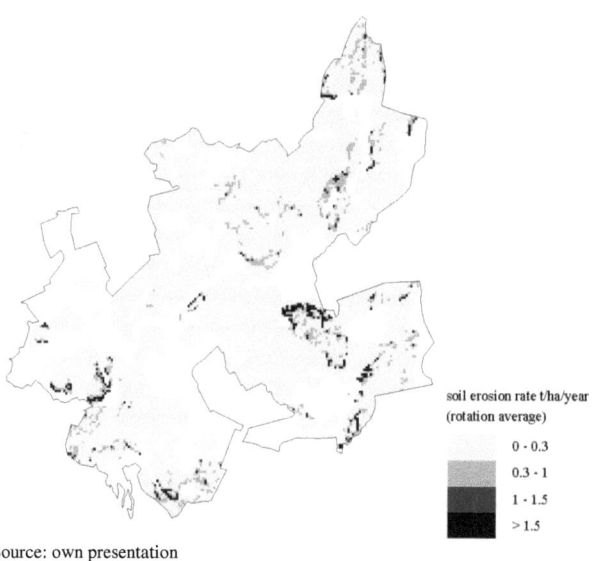

Source: own presentation

Figure 35: Average soil erosion risk for the region under targeted incentive conditions

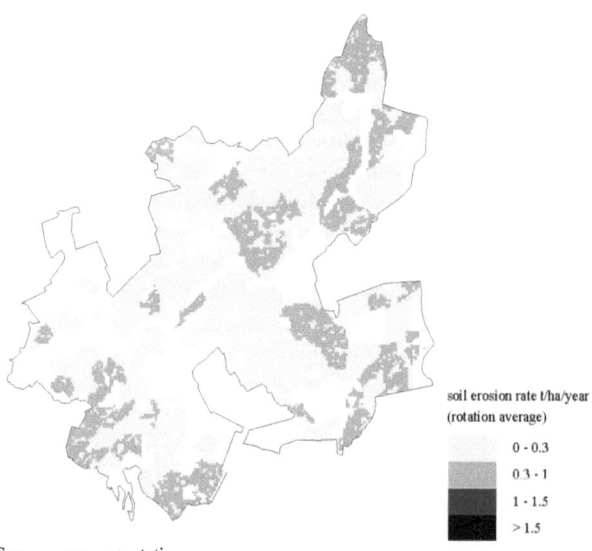

Source: own presentation

Figure 36: Average soil erosion risk for the region under conditions of row crop restrictions

7.8.5 Changes in livestock production and labour needs

The conditions in the scenarios also have effects on the livestock production and labour needs. The dairy sector was only moderately affected, i.e. the amount of milking cows kept in the region stayed the same for all scenarios except for a small increase in the restriction scenario. The fattening bull sector, which was completely given up under the CAP2013 conditions, did not receive any incentive through the soil conservation scenarios to start producing again. As for pork production, the scenarios showed diverse results: Under the CAP2013 conditions, almost 16,000 pigs were produced in the region. The incentive scenarios changed the opportunity costs for this livestock system in such a way that pork production became less profitable. In the row crop restriction scenario, the production of pork increased by almost 5,000 units as compared to the CAP2013 scenario (see Table 51).

Table 51: Animal numbers in the scenarios

Scenario		CAP2013	Untargeted Incentives	Targeted Incentives	Row crop restrictions
Cows	numbers	1,973	1,973	1,973	1,995
Fattening bulls	numbers	0	0	0	0
Pigs	numbers	15,947	6,007	5,446	20,223

Source: own calculations

The effects of the scenarios on the labour demand in the region also varied. Table 52 shows the labour demand under the different scenarios itemized by plant and livestock production. All reduced tillage scenarios showed a decrease in labour demand compared to the CAP scenario, due to the reduction in pork production, while the row crop restriction option provided more labour opportunities. However, it is not possible to conclude that agricultural labour demand could be directly increased by soil conservation policies. The effects on livestock production are not related directly to the policy, but rather based on the opportunity costs of pork production in these examples. Pork prices vary for many reasons, so the effect of such conservation policies would instantly be counteracted.

Table 52: Labour demand in the scenarios

		CAP2013	Untargeted Incentives	Targeted Incentives	Row crop restrictions
Labour plant production	numbers	27	27	26	28
Labour livestock production	numbers	51	43	43	54

Source: own calculations

7.9 Conclusions

7.9.1 Policy options

The effects on soil erosion risk and the economic situation of a region was shown in the preceding chapters using examples of selected soil conservation policies. Based on the conclusion of market failure, a policy intervention was justified. A cost-effectiveness-analysis was applied following a safe minimum standard approach so that an efficient solution to reduce soil erosion risk may be found.

The examples showed that policies based on reduced tillage incentives can positively influence the erosion situation in a region. However, given the voluntary nature of the incentive policies, the actual uptake of such programmes is subject to the surrounding conditions (e.g. relative prices, attitudes of farmers), which might even cause adverse effects on such policies. The row crop restriction option also achieved similar erosion risk reductions.

When the costs of policies were examined, the restriction option was proven to be highly effective in terms of on-farm costs, while budget costs could not be considered in this framework. Incentive based options showed high total costs due to both budget and on-farm costs.

When looking specifically at the cost-effectiveness of these policies, row crop restriction seems to be advantageous over the other options. However, since the costs of policies in this framework were restricted to on-farm and budget costs (representing the sum of payments for certain measures), the result was therefore biased in favour of row crop restriction. Compensation payment to farmers for losses through such a policy would also increase the budget costs for the row crop restriction policy. When the on-farm costs induced by the policies were compared to the benchmark scenario of the social planner, the row crop restriction scenario came very close to this result. However, the changes induced by each scenario were different. While the row crop restriction banned all the high erosion crops from high risk spots, the social planner option only shifted certain crops but still kept certain highly profitable crops on the high risk areas.

Note that the policy scenarios are not driven by the goal "soil erosion risk reduction" but by the maximization of the rent that could be achieved through the reduced tillage incentive or by the minimization of the negative effects of a crop restriction on the total gross margin. As a result, even adverse effects that were not intended by the design of a policy can occur.

This outcome shows a dilemma that holds true for many agri-environmental policies: if a policy cannot specifically target a certain improvement of an environmental situation, the policy maker has to rely on a correlation between an agricultural measure and the environmental effect. Therefore, all action-oriented policy instruments face the risk of not having direct influence on the environmental objective, but only influence on a specific agricultural measure, which is only to a certain extent correlated with the improvement of the environmental situation. This effect could turn the policy instrument into a policy failure, depending on the level of desired outcome achieved.

The high budget costs for the incentive based policies allowed for the large windfall gains for reduced tillage practices that were already in practice under the CAP2013 scenario. Additionally, the effect of reduced tillage practices on erosion reduction is actually rather limited compared to crop change. Therefore, almost three quarter of the area had to be made eligible in order to meet the envisaged erosion reduction level which resulted in the extremely high budget costs. Furthermore, there were crop types used that were rather unsuitable for reduced tillage.

The difference between the budget and farm costs in the incentive policies shows to a certain extent, the overcompensation from the incentive payments. In the single field example, incentives are truly needed to compensate for the use of less profitable cropping practices on the single field. However, a single field example does not take into account a farm's ability to compensate for the loss from conservation practices with its other fields. Therefore, windfall gains are more or less unavoidable.

Based on the cost-effectiveness criteria and on the assumptions of the modelling framework row crop restriction was shown to be the preferable policy among the tested policy options even if compensation payments for the on-farm costs would be paid.

However, in reality additional costs will arise for all policies. It can be expected that control costs will arise for all three options. The proper execution of reduced tillage has to be monitored for the incentive options, while the row crop restriction policy is only dependent on the appropriate control mechanism of such a regulation. Such costs can affect the overall effectiveness in terms of money spent per ton of reduced erosion and therefore need to be considered in the analysis of policies. The relevance of such costs that go beyond the basic on-farm and budget costs will be discussed in Chapter 8.

Further conclusions on the theoretical concept are drawn in Chapter 9.1.

7.9.2 The modelling system

The application of the chosen modelling system helped show the resulting on-farm costs and budget costs for all policy options based on the underlying assumptions. It was able to show the effects of both the soil conservation policies and the effects of the general policy changes (from the conditions of Agenda 2000 to the new CAP reform).

The indicators chosen in this approach were helpful in the analysis of the effects of policy changes. The basic indicators "total gross margin" and "soil erosion risk levels" provided general information on the conditions following the implementation of new policies. The derived indicator "cost-effectiveness of soil conservation policies" shows clearly which policy spends the least money for the reduction of soil erosion in a region, where these costs occur and the form they take (either as on-farm or as budget costs).

The availability of the complete range of cropping practices for each crop in the model is very important. If certain crops are not defined with reduce tillage options, the model results can be biased.

Reduced tillage practices are in general more profitable in terms of lower labour costs. High incentives for such tillage systems create unrealistic combinations in the resulting model solution. Given a soil conservation measure with slightly higher costs (e.g. zero tillage in combination with specific machinery) than a standard measure, the model can be expected to increase the area of reduced tillage from very low to much higher levels.

The degree of detail in the modelling system allows for the specific analysis of the crop shares grown, the spatial distribution of certain crops as well as the indirect effects on livestock production and labour demand. Therefore, the model approach can help as a decision support for policy makers.

The model describes changes in behaviour based only on the assumption of profit maximization, while other motives of an entrepreneur are not taken into account. Risk minimization, social influences within the community of the region and personal preferences can change the reaction of farmers in comparison to the model results.

Furthermore, the model was not based on a dynamic approach, so the financing costs of investments in new machinery for soil conservation measures are not modelled as clearly as in a dynamic model. In order to avoid the strong influence of these effects, only measures that could be realised with standard farm equipment were modelled.

8 Transaction costs, property rights and soil conservation

8.1 Background

In the preceding chapter, the implementation of soil conservation programmes was analysed through cost-effectiveness with regards to on-farm costs and direct budget costs of incentive payments. The implementation of these agri-environmental policies raised further questions regarding which instrument is preferable from a transaction costs and property rights perspective.

In this chapter, such questions are discussed so that a synthesis of the modelling results and the application of a wider economic framework can be found.

8.2 The scope of analysis

In the case of governmental intervention justified by market failure, governments have to decide which policy should be promoted to meet society's needs for soil protection, be that more extension services or the financial support of certain cropping systems. Furthermore, they have to decide, whether the programmes should be targeted at certain sensitive areas or if the programmes should apply to every farmer within a country. Regardless of the decision, each choice brings further costs with it that consist not only of the money paid to the farmer for compensation or as an incentive. There are also administrative costs, i.e. the programmes have to be implemented, controlled and fraud has to be prosecuted (McCann et al. 2005). All these hidden costs are summarised under the term transaction costs that was initially characterized by Coase (1937) as the main determinant for allocative decisions within economic systems. In the same context, it is important to know the distribution of the property rights for the relevant goods, since their allocation is a crucial point in whether a programme will be successfully accepted (Challen 2000).

Transactions costs and property rights are terms commonly used in the context of New Institutional Economy. This chapter gives a brief overview of some of the economic theories from this branch of economics that can be used to explain the problems that evolve from the design and implementation of more sustainable policies. Furthermore, the use of these theories within this study is described in the examples of soil conservation policies.

The emphasis is on the evaluation of the administrative part of transaction costs, which is often neglected in the process of designing agro-environmental programmes (Falconer and Whitby 2000). Given the constraints in time, the optimal allocation of property rights cannot be evaluated within this study. This is also because fundamental changes in property rights without compensation require a lot of time for the legal preparation within a democratic system and will face high resistance among the affected groups: either as a result of rent-seeking processes (Buchanan et al.

1980) or as an outcome of the given distribution of political power (Knight 1992). Within the short-term process of policy-making an uncompensated change of property rights is a difficult task. Therefore, for the following discussions it is assumed that decision makers develop policies based on a given set of property rights, using moderate tools such as legal restrictions accompanied by compensations or financial incentives for certain measures to induce behaviour change instead of a strict redistribution without compensation through legislative orders. Hence, the focus of this study is to find out, how the total costs of soil conservation programmes (i.e. including transaction costs) can be minimised for specific scenarios of reduced erosion levels.

8.3 A brief overview on New Institutional Economics

8.3.1 Transaction costs

Transaction costs are the basic subject in most theories of New Institutional Economics. Transaction costs determine the behaviour in or between organisations.

Some definitions for transaction costs are (all citations from Hubbard 1997):

- the costs of arranging a contract ex ante and monitoring and enforcing it ex post (Matthews 1986)
- costs of running the economic system (Arrow 1969)
- the economic equivalent of friction in physical systems (Williamson 1985).

A very general, but precise definition was given by Allen (1991): "(…) I define transaction costs as the cost of establishing and maintaining property rights. This definition illustrates that these costs arise out of more than information costs, that they are not just like taxes, and that they are necessary to explain any distribution of property rights." Furthermore he stated that "incomplete property rights and transaction costs are two sides of the same coin."

8.3.2 Property Rights

Characteristics of a property right are the right to manage a defined object, the right to receive income from using the object and the right to alienate or sell the object (Scott 1989a; Scott 1989b; cited in Challen 2000).

Bromley (1989) used the term property right very broadly and linked it to the term institution in the economic sense: property rights do not only define the ownership of a person to an object. According to his definition, property rights include use rights, exchange rights, distribution entitlements, management systems and systems of authority and enforcement. Property rights incorporate all the institutional rules that govern the ties between an individual, the society and a certain object.

Challen (2000, p. 15) described "property rights as the subset of institutions for the regulation of behaviour and social interactions with respect to objects of value. In an institutional context, property refers to the rules of behaviour rather than the object." Bromley (1989; cited in Challen 2000), also stressed the meaning of property rights more as the "social relation amongst individuals within a society than a relationship between an individual and a particular object of value."

The economic definition of institutions as "the rules of the game" (North 1990) or "social rights and obligations" (Hubbard 1997) comprises also the term "property rights" as a certain form of institution. Transaction costs are used in this context to describe the allocation costs of property rights.

One of the objectives of NIE is to analyse the social rights and obligations (institutions). The central point is the draw up, monitoring and enforcing of contracts. All this is reflected in transaction costs as the extent of imperfect information (Hubbard 1997).

The time dimension is important in NIE, since institutions can only be appropriate for a certain time, but will change under the pressure of different scarcities. In short, NIE attempts to show why neoclassical economics often fails to explain real world economics (Hubbard 1997).

The given distribution of property rights is of some relevance, since transaction costs can vary highly if property rights on certain attributes can be transferred to or shared between different agents (Lippert 1999; Lippert 2005).

In the context of soil conservation, the property rights on the land used by farmers play an important role. If farmers own the full set of property rights of the their land, i.e. the right to use it, sell it and even destroy it, then all the efforts to limit soil erosion would have to be on a voluntary basis, aimed at the limited purchase of certain parts of the property rights. However, these property rights are usually limited by public interest represented by jurisdiction or based on agreements between land users. Ciriacy-Wantrup and Bishop examined the institutional performance of common property institutions and concluded that there are many examples where these tools could serve adequately in the management of public property. However, they warned against the dangers of neglecting the distinction between common property institutions and the absence of rights to property (Ciriacy-Wantrup and Bishop 1975).

Furthermore, if the use of soil harms other members of society, it is very likely that a negotiation process would be started, where the benefit the farmer has from using the soil and the harmful effects the neighbours will suffer from are discussed like in the ideal case described by Coase (1960). Under real world conditions this negotiation process would take time and may not be as clear cut as Coase proposed. In addition, society could in fact cut the full range of property rights by dispossession, should farmers jeopardise the long term food security of a country.

One of the main conclusions of Coase's work is that property rights have to be clearly defined and allocated, be it for the pollutant or the person suffering from pollution. If the interest of one of the parties is important enough a process would be started so that a solution where both are better off could be reached (see Chapter 4.4.2.2).

8.3.3 An overview of the basic theories

The following paragraphs provide a brief overview of some basic theories of New Institutional Economics (Hubbard 1997):

- Theory of the firm (transaction costs):

Based on the article "The nature of the firm" by Coase (1937), transaction costs determine whether a firm or the market is the better institution for organizing economic behaviour. High transaction costs within a market due to high risks, uncertain conditions and low trust between market partners, is more likely to result in a higher percentage of resource allocations within a firm. Later, the article "The Problem of Social Cost" (Coase 1960) introduced transaction costs in a wider context of property rights, compensation and negotiations in the attempt to achieve a pareto-optimal solution (see Chapter 4.4.2.2).

- Theory of the markets (imperfect information):

Focussing more on the organisation of markets than on the relations between markets and firms, this theory attempts to find explanations for non-market solutions between economic organisations. Due to imperfect, asymmetric information and malfunctioning markets, non-market solutions are chosen for the co-ordination of economic transfers. Markets cannot develop within very weakly manifested institutions such as in developing economies or when the risk of economic action is not sufficiently covered by the price (Stiglitz 1974; Stiglitz 1986).

- Theory of politics (institutions used to favour interest groups):

This theory outlines the influence of politics on the distribution of property rights. Small but powerful interest groups of producers are able to exert more on economic decisions than e.g. consumers. This "dark side" of pure economy allows the initial distribution of wealth and power underlying allocation decisions ((Bates 1989; cited in Hubbard 1997)).

White (1993) defined four sources of power in the economic performance of a society as: state, collective action by market actors, market structure and socially embedded power through birth or status.

- Theory of history (institutional change):

This theory analyses the influence of society's economic and institutional history on the status quo of institutions and the performance of economic behaviour. Societies evolve new rules to reduce transaction costs as they move along the path of development. When production costs decrease, transaction costs would rise as a result of more risk and uncertainty. New rules can reduce transaction costs just as technical progress decreases the costs of production. According to this theory, the actual situation of the economy is influenced by historically set path effects (North 1990).

8.4 Transaction costs in the context of soil conservation programmes

8.4.1 Transaction costs in policy evaluation

The act of measuring transaction costs in soil conservation programmes requires a definition of transaction costs that is more suitable for the evaluation of policies. A definition that leads further in the context of soil conservation programmes was given by Thompson (1998): "Institutional transaction costs (ITC) include the costs of enacting a policy by a legislature, and the costs of implementing and enforcing that policy by administrative agencies and the courts." Here, the institutional character of transaction costs becomes apparent. Transaction costs are always related to institutional arrangements, but in the Thompson example, institutions were defined more in the organisational meaning, focussed on the political context of transaction costs. A more precise description of this type of transaction costs are setup and administrative costs.

The following chapters describe the design of a framework for estimating transaction costs of soil conservation programmes.

8.4.2 Boundary issues

The first question leads to the institutional boundaries of the analysed policies (McCann and Easter 2004): How far does a policy change influence the set-up of property rights?

Figure 37 describes the different institutional areas that may be relevant in the measurement of transaction costs.

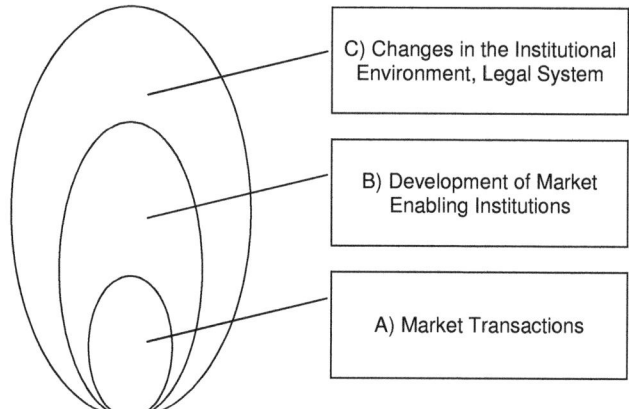

Source: McCann and Easter 2004; own presentation

Figure 37: Boundary issues related to transaction costs stemming from market transactions, market enabling institutions and changes in the institutional environment and legal system

The area 'A' corresponds to the pure "market" transactions for a policy, i.e. the monetary exchange of taxes or subsidies and the resulting change in production practices. The authors claimed that focussing only on these costs would be too limiting. Most policy implementations are accompanied by the development of specific institutions such as regulations, controlling agencies and set-up costs in the political process. Therefore, in order to measure the transaction costs of a policy, at least both cost types (areas A and B) would have to be considered. If the policy requires change in the institutional arrangements or a new legal system, the field of transaction costs would be even wider (area C). In such cases, costs for all the areas would have to be analysed.

8.4.3 Stakeholders and transaction costs

Given the institutional boundaries, the analysed soil conservation policies often do affect a wide scope of stakeholders. Thompson (1998) provided a list of actors affected by transaction costs as follow:

- legislators,
- interest groups,
- administrative agencies,
- courts and
- the private individuals regulated by the policy.

An overview of the involved agents' functions and some examples are shown in Table 53.

Table 53: Functions and examples of agents involved in the implementation of a soil conservation policy

Function of agent	Agent(s)
Legislators	EU-Council
	German Federal Government (Bundestag)
	Federal state governments (Landtag)
Interest groups	Farmers' Union
	Environmentalists
	Soil science experts
Administrative agencies	EU-Commission
	Soil agency
	Regional enforcement agencies
Courts	Constitutional Court
The private individuals regulated by the policy	farmers
	persons affected by off site damages caused by erosion (i.e. residents along farmland, fishermen, power stations and recreational users of water courses)

Source: own presentation based on Thompson 1998

The role of the different stakeholders will be further discussed in Chapter 8.5.2

8.4.4 Forms of transaction costs

For the analysis of transaction costs in agri-environmental policies (i.e. soil conservation policies), it is helpful to categorize them into specific groups. Furthermore, the occurring costs can be assigned for different purposes. McCann et al. (2005) expanded on a framework of the different forms of transaction costs based on the initial design by Thompson (1998). This Institutional Transaction Costs framework (ITC) describes also behavioural assumptions of all actors involved in the policy making process. It gives a guideline for setting up policy analyses of different policy options.

Table 54: Typology of transaction costs associated with public policies and parties incurring costs

Type of transaction costs		Incurred by		
Thompson 1998	McCann et al. 2005	Legislature/ courts	Agencies	Stakeholders
	Research and information	+	++	+
Enactment	Enactment or litigation	++	+	++
Implementation	Design and implementation	o	++	+
Compliance		o	o	++
	Support and administration	o	++	+
	Contracting	o	+	++
Detection	Monitoring/detection	o	++	+
Prosecution	Prosecution/enforcement	+	++	+

(o) Negligible transaction costs; (+) low transaction costs; (++) high transaction costs.
Source: McCann et al. 2005, Thompson 1998; own presentation

Table 54 combines the categories from the frameworks provided by McCann et al. (2005) and Thompson (1998). The costs describe types of transaction costs accrued on the different stages of the development and implementation of policies. In the following explanations, an approach based on Thompson's theory will be used, for the categories are simpler and more helpful for the rough distinction of cost types.

It is difficult to assign research and information to a specific policy. Therefore, according to the authors, only research that is directly related to a policy design should be accounted for. This aspect however, will not be further analysed in this study.

The enactment costs can be taken as a lump sum, for the number of Parliament members does not change with policy changes. The same might be true for lobby groups. However, these costs are beyond the scope of this study.

Implementation costs involve the determination of both its goals and means through an administrative agency. Such costs vary highly depending on the precision and site-specificity of the chosen instruments.

Thompson (1998) proposed a group of compliance costs for stakeholders that are part of enactment and contracting costs in the McCann categories. The compliance costs of farmers related to a certain policy scheme are defined as the additional organisational effort in labour time for participating in a programme or complying with a regulation (e.g. applying for programmes, information gathering on regulations).

Detection and monitoring activities provide information on whether participants follow the new regulations of the policy and comply with the contracted management agreements.

Prosecution costs arise when agencies have to enforce a policy i.e. in the case of the violation of policy regulations by the affected stakeholders.

Falconer et al. (2001) provided a more detailed categorisation of transaction costs adapted for the voluntary environmental stewardship schemes in Great Britain, which are usually site specific and negotiated individually (i.e. ESA – Environmentally Sensitive Areas). Table 55 shows the level transaction costs are expected to occur, the actors that will be affected and whether the transaction costs are dependent on the size of the supported area or the number of participants. This framework can also be transferred to categorize the different options of voluntary soil conservation programmes.

Chapter 8 – Transaction costs, property rights and soil conservation

Table 55: Categories of transactional costs incurred in the implementation of voluntary schemes based on compensated management agreements and cost incidence

Main Category	Sub Category	State Agency Costs		Participant costs	
		Fixed at the level of the scheme	Variable with no. of participants	Fixed at the level of Participant	Variable, e.g. with hectares entered
Information	Survey of designated area	X			
	Designation of area and prescription design	X			
	Re-design/re-notification of prescriptions	X			
Contracting	Promotion of scheme to farmers		X	X	
	Negotiation between organisation and farmer	X	X	X	X
	Administration of contract (including making payments to farmers)		X	X	
Policing	Enforcement of farmer compliance		X	X	X
Evaluation	Environmental monitoring and scheme evaluation	X			

Source: Falconer et al. 2001; own presentation

The first step at the implementation level is gathering information on the area and the suitable measures for conserving certain attributes in this area. The increase in knowledge regarding the area or the environmental good could create the need for redesigning the conservation measures. This part creates only more costs on the government side.

When contracts are negotiated or placed with farmers, the possible participants have to be informed, the payments have to be negotiated (in case it is a very individual contracting scheme) and the whole programme has to be administrated. Even voluntary programmes with a fixed payment require a certain amount of promotion, since the participation depends on information provided to the farmers. At this stage, costs can arise on both government and farm side.

The policing category comprises costs that arise from enforcing farmer compliance. The state agency faces costs that vary with the number of participants. A participant on the other hand, might have variable compliance costs depending on the amount of land entered into the scheme.

The evaluation of the environmental effects of conservation programmes creates costs only on the governmental level. According to the authors, these costs are fixed at the level of the scheme. This means that a large number of participants with only small shares of land in the scheme would create higher monitoring costs than monitoring the effects on a piece of land that is owned by one individual farmer.

Both approaches use similar categories with overlaps in certain parts. The analysis in this study will be based on the categories by McCann et al. (2005) and Thompson (1998).

Another issue in the policy developing process is time. Transaction costs do not occur at the same time and constantly over time (McCann et al. 2005). Some policies might have high set-up costs but low costs after implementation, while other policies cause a steady flow of costs (e.g. permanent control costs).

8.4.5 Recent attempts in measuring transaction costs of environmental policies

There had only been a few attempts to categorize and measure transaction costs of agri-environmental programmes. Falconer et al. (2001) and Falconer and Whitby (1999) estimated the transaction costs of implementing countryside stewardship programmes in the United Kingdom and in Europe. Westra et al. (2002) analysed the transaction costs of policy options for phosphorus reduction in the watershed of Minnesota River. McCann and Easter (2004) proposed two ways of measuring transaction costs of environmental programmes: either 1) by surveys or interviews to estimate transaction costs or 2) based on government expenditure reports.

Surveys or interviews are time consuming and thus costly (McCann and Easter 1999b) but they make it possible to obtain information on the full range of relevant costs and implicit as well as explicit costs.

Transaction cost measurement **based on government expenditure reports**, (e.g. Falconer et al. 2001, Falconer and Whitby 2000, McCann and Easter 2000), has the advantage of representing actual expenditures and not requiring surveys or interviews. However, the authors did also list a number of disadvantages (McCann and Easter 2004): governmental data do not completely cover the costs desired by researchers or they cannot by assigned to specific policies. The agencies have to cooperate with the research project and spend effort to get data together. Data can be confidential and/or is only available after the policy has been implemented.

When transaction costs are measured a trade-off exists between precision and measurement costs: if only available data are used and relevant and difficult to gather data are neglected, the analysis would fail (McCann et al. 2005). As the authors stated, "an initial screening across policy instruments, rough "orders of magnitude" may be good enough and would represent an improvement over current practice" (McCann et al. 2005), p.521).

A final problem arises from the fact that either implicit or explicit costs occur: family labour has differing opportunity costs within seasons and according to the personal interest of a farmer. The reallocation of a staff that is already working in an environmental agency causes implicit costs while hiring new personal creates explicit expenses (McCann et al. 2005).

If transaction costs cannot be measured, a description of the involved cost types can at least help in improving the design of a policy (McCann and Easter 2004).

In general, the above mentioned rough orders of magnitude help give one an idea of the types of costs that can occur with the implementation of new conservation programmes. Therefore, this study will only aim at a qualitative, cardinal evaluation of the involved transaction costs. The result of this approach is not a monetary assessment of the transaction costs of a policy. Due to the difficulty in measuring transaction costs, a more descriptive approach will be followed to provide information on the feasibility of soil conservation programmes.

8.4.6 Suitable reference values for soil conservation policies

The measuring and analysing of transaction costs require first of all a specific reference value for the comparison of different options. Some authors suggested comparing the share of transaction costs to the total spending of a policy (McCann et al. 2005).

However, this is only possible if the policies are comparable in terms of compensation levels and policy design. For the policy options compared in this study, this dilemma is obvious for budget and on-farm costs. If the incentive based programme is compared to the regulation option with no direct payments to farmers, the regulation approach would consist totally of transaction costs according to the above definition of institutional transaction costs. Nevertheless, the regulation approach could be cost-efficient in terms of total costs of the policy.

Besides, if a low TC share of the total budget is assumed to be efficient, this would imply that the higher the payments to farmers are, the more efficient this policy would be. Again, this is only true when similar policies with equal budgets are compared.

Other indicators such as transaction costs per single contract or per hectare of contracted plots could face similar difficulties if the type of compared policy is too diverse. If transaction costs of different policy types are to be described, the ideal indicator is to relate the transaction costs to an environmental result. In the case of soil conservation, a certain reduced level of erosion per hectare would be the most appropriate indicator.

Excursus: The influence of attitudes

It must not be forgotten that attitudes towards conservation programmes play an important role in the participation of farmers in environmental conservation programmes (Drake et al. 1999; Falconer 2000; McCann and Easter 1999a). More detailed information regarding the reason for nature or soil conservation can influence attitudes towards these programmes. Negative attitudes towards these programmes are often caused by a lack of information. However, the distribution of information is not a costless effort either. There is a trade-off between lowering the uptake costs for farmers and raising costs on the administration side through the provision of information.

However, attitudes were not analysed within this study. Attitudes can vary among farmers and explain participation in environmental programmes individually. For a regional modelling approach, attitudes are unlikely to be describable.

This is in line with Falconer and Whitby (2000) who stated that the overall efficiency of a policy should be the aim of the analysis, i.e. the total costs per indicator value should be looked at, which include transaction costs and payments to the farmers.

For this study, the potential amount of erosion of a region, which is derived by the bio-economic modeling in the previous chapters, or the potential average soil erosion per hectare serve as appropriate indicators for soil erosion for the comparison of the different policy options.

8.4.7 Attributes of an environmental good and transaction costs

Each environmental good can show characteristics that can have an influence on the potential transaction costs and the appropriate instrument for its promotion. Weersink et al. (1998) outlined that there is no overall first best instrument for the promotion of all environmental goods that are demanded by society. An appropriate policy minimises (the sum of) the environmental costs of the external effects of agricultural production (residuals), as well as the abatement costs of the producers and the administrative costs for regulation, monitoring and enforcing compliance.

Falconer et al. (2001) related the appropriateness of an instrument to the variability and heterogeneity of the participating farmers including the properties of their farms (in terms of their opportunity costs) and to the variability of the environmental good. The higher the variability of costs and goods, the more individually negotiated are the agreements (Table 56).

Table 56: Appropriate instruments depending on the variability of the environmental good and producer/Production type

		Variability of producer (in terms of agricultural opportunity costs)	
		Homogeneous	Heterogeneous
Variability of agri-environmental good	Homogenous	Standard contracts and payments for specified goods and services	Auctions
	Heterogeneous	Site-specific management agreements and payments	

Source: Falconer et al. 2001

Falconer and Whitby (1999) gave a more detailed classification of policy options and their administrative costs to promote agri-environmental goods (Table 57). It was stressed that depending on the type of policy approach chosen, different administrative costs will occur. However, the general framework from rather voluntary to more restricting measures, changing the distribution of property rights fundamentally, is still the same.

Table 57: Policy approaches and administrative costs

	Information, set-up, promotion	Contracting	Policing
Persuasion and Advice	X		
Regulation	X		
Market mechanisms (e.g. taxes)	X		
Tradable permit schemes	X	X	X
Voluntary management agreements	X	X	X
Public purchase of land	X	X	

Source: Falconer and Whitby 1999

8.4.8 Research on transaction costs in environmental policies

The following paragraphs show examples for quantifying transaction costs in environmental policies and list arguments brought up in the discussion of environmental programmes from a transaction cost perspective.

An attempt to calculate transaction costs was made with the comparison of the transaction costs of organic farming to those of a set of voluntary standard environmental programmes with comparable effects (Hagedorn et al. 2004; Tiemann et al. 2005). Organic farming was proven to be a policy option that decreases transaction costs compared to single measures when administrative costs were analysed.

Falconer and Saunders (2002) compared the scheme-related transaction costs of individually negotiated and standard management agreements in a long term nature conservation scheme for sites of special scientific interest in the North of England. They showed that in the specific case of these programmes, individually negotiated agreements bore less transaction costs than standard management agreements.

A study was performed to measure the magnitude of transaction costs associated with policies to reduce agricultural nonpoint source pollution to specified levels in the Minnesota River (McCann and Easter 1998; McCann and Easter 1999b). Interviews with staff from governmental agencies were conducted to estimate the associated transaction costs. The results showed that the tax policy on phosphate fertilizers had the lowest transaction costs, followed by educational programs on best management practices, the requirement for conservation tillage on all cropped land, and the expansion of a permanent conservation easement program. The requirement for conservation tillage on all cropped land was less cost-effective due to the high control costs of the command and control approach (McCann and Easter 1999b).

Westra et al. (2002) reported lower transaction costs for targeted as opposed to non-targeted policy options for phosphorus reduction in the watershed of Minnesota River. With transaction costs taken into account, the costs of a targeted programme could be outweighed by a higher cost-effectiveness (Westra et al. 2002).

In general, some decision rules for the choice between voluntary and regulation approaches should be considered:

High heterogeneity of an environmental good justifies extra efforts for spatial targeted programmes (Falconer et al. 2001).

The severity of a soil erosion problem can be the trigger for one of the options. If soil degradation is developing at an alarmingly high rate, a command and control strategy would be preferred by governments and uncertain voluntary adoption may be avoided (Oates and Portney 2001). However, high rates of soil degradation could also increase the awareness of land users and support voluntary adoption of soil conservation measures, as stakeholders would be more involved in the resource problem (Ostrom 1991). Additionally, the threat of implementing mandatory regulations could also force land users to adopt voluntary programmes in order to avoid more drastic measures from a command-and-control policy (Segerson and Miceli 1998).

Latacz-Lohmann (2001) underlined that voluntary approaches create the notion of fairness. The award of windfall gains from conservation programmes to participants with low compliance costs can be avoided through a regulation policy with least cost compensation payments.

Even though property rights were assumed to be similar for all policies (if a compensation payment was considered in the regulation approach), the political price of changing the institutional arrangement of soil conservation attempts can vary between the options analysed. A mandatory regulation usually faces higher resistance compared to a voluntary approach (Latacz-Lohmann 2001). Therefore, the feasible option becomes clearer through the political process of discussion and testing of options by the involved stakeholders.

Altogether, these results show that transaction costs depend on several factors such as targetedness, economies of scale, specificity of the programmes and the selected instrument itself. These factors are also reflected in the following analysis.

8.5 *Qualitative analysis of transaction costs of soil conservation policies*

The aim of this chapter is to derive the magnitude of transaction costs for the administrative part of policies and to point out the differences between the three policy options analysed.

In contrast to the on-farm costs, the estimated transaction costs related to the different scenarios are much more of a qualitative nature. Due to the different levels of detail, measurability and quality, it would not be accurate to sum up the results of the modelling approach with the following qualitative considerations.

In the following, the different aspects from Chapter 8.4 are discussed using the specific example of the above analysed soil conservation policies.

8.5.1 Boundary issues

The soil conservation policies analysed in this study were based on a given set of property rights, i.e. it is assumed that the right to use soil is either claimed directly by the farmers (in the incentive option) or purchased partly by offering incentives. Or, in the case of a regulation, the change in property rights must be compensated to the owners of the soil. Therefore, using the terminology of chapter 8.4.1, the set-up costs of specific institutions such as regulations and controlling agencies must be considered. All policies must be based on a legal regulation:

- In the case of an incentive approach, the agency must be authorised to offer incentives for specific measures. This usually needs to be based on EU-regulations. Then, the regulations are further implemented on the national (and federal state) level.
- In the case of specific land use restrictions, they can be either based on national or regional planning decisions, or, be part of an EU-wide framework directive (Commission of the European Communities 2006). The same is true for a targeted approach that needs to be based on a legal act for the selection of the regulation area.

As a result, the institutional boundaries for all policies need legal regulations (see Figure 37, p.140, area A and B) and therefore, can be assumed to be similar except for targeted approaches, which require greater efforts.

8.5.2 Agents involved in the policy making process

Transaction costs affect different actors within the process of developing an agri-environmental programme. The following chapter highlights the actors that could be possibly involved and their assumed behaviour in such a policy making process.

Most agri-environmental programmes are based on Council Regulation (EC) No 1698/2005 (European Council 2008a), which regulates and approves the design of these policies. Furthermore, since the adoption of the "Thematic Strategy for Soil Protection" (Commission of the European Communities 2006), soil conservation will be increasingly addressed on the European level[14].

On the national level, the relevant legislators are the German Federal Government (*Bundestag*) that adopts national laws (e.g. the German Soil conservation act) and the governments of the German federal states (*Laender*) that have to implement national law on the federal state level by means of an enactment. The way the Federal Act will be implemented can vary between the federal states.

The federal state governments (*Laender*) are likely to formulate the enactment of the EU-guidelines in a way that meets the interests of certain interests groups (Latacz-Lohmann 2001). The relevant interest groups involved in soil conservation aspects are farmer unions, environmentalists and

---

[14] Even though the proposal of a draft directive for soil protection was not accepted by the European Council (European Council 2008b).

experts in soil science. The pressure from some interest groups can influence the administrative agencies (federal, federal state and local agricultural and environmental agencies) to formulate the enactment in such a way that EU funds are likely to be used to minimise farmers' costs and externalize most of the incurred costs to the EU budget (Latacz-Lohmann 2001). The formulation of the German Soil Protection Act was influenced on the federal level by interest groups (in this case the farmers' lobby) in such a way that "good technical practice", a rather diffuse definition of proper land use was assigned to be the appropriate way to prevent soil erosion (Landel et al. 1998) (see also Chapter 3.3.2). Even independent expert commissions have their individual, subjective interests, since it often is their goal to promote a maximum level of soil conservation. To some extent, this group assigns an intrinsic value to soils that is not derived from actual, potential or future use values (Cicchetti and Wilde 1992). However, it is the task of policy-making institutions to facilitate the differing interests of society. For the discussion on values reference is made to the works of Dabbert (1994), Navrud (2000), and Pearce (1993).

On the juridical level, the Constitutional Court can be involved if the legislation process is considered inconsistent with the German constitution, thereby impairing the property rights of land owners. Private individuals regulated by the policy are the farmers, but also everyone that suffers from the degradation of soils and off-site damages caused by erosion (i.e. residents along farmland, fishermen, power stations and recreational users of water courses).

The above considerations can affect all of the analysed policies in a different extent. Incentives will face of course less opposition among farmers than restrictions, especially if windfall gains are possible. However, in times of limited budget and increasing justification pressures against subsidies, taxpayers will favour restrictions.

A further distinction can be made for targeted policies: in this case, the stakeholders affected are divided into two groups: one group benefits from or suffers under the new policy while the other group is not affected, but might feel disadvantaged because it cannot participate in a soil conservation programme. In this case aspects of fairness and equity become important. Latacz-Lohmann (2001) illustrated that "farmers participating in an agri-environmental incentive scheme may find it "fairer" if all participants receive the same payment for the same level of commitment (egalitarianism), while it would be more cost-effective to offer different farmers different payments according to their individual compliance costs (proportionality)."

This example is also valid in the extreme case of total exclusion by means of a targeted programme. Farmers with fields in a non soil erosion risk area cannot offer the service of erosion reduction, and therefore cannot expect any payment for it. This aspect needs to be considered when spatially targeted programmes are taken into account.

8.5.3 Forms of transaction costs

In order to find an efficient policy option, the forms of transaction costs have to be evaluated and compared. Given the difficulties in measuring transaction costs, a description of the relevant transaction costs in a qualitative, comparing way provides an overview of the expected costs. Using the Thompson classification of Table 54, the procedure is described below (see Table 58).

Table 58: Estimation of transaction costs types for the policy options in this study

Cost type	Evaluation approach in this study
Enactment costs	Assumed to be equal for all policy options
	For spatial targetedness higher enactment costs can occur.
Implementation costs	Considerations based on Table 57 plus additional efforts for targeted approaches
Compliance costs	Administrative costs through information gathering, increased organisational tasks, monitoring obligations on farm level
Detection costs	Assumed to be similar
Prosecution costs	Assumed to be similar

Source: own considerations; categories based on Thompson 1998

The enactment costs are assumed to be equal for all policy options given the similar legislation process for agri-environmental incentive programmes and legal regulation. However, depending on the spatial targetedness higher enactment costs can occur. For the implementation costs, the following considerations are based on the cost levels stated in Table 57 (see Falconer and Whitby 1999), which lead to additional costs for both the contracting and policing of voluntary management agreements (incentive options) and targeted approaches (targeted Incentives and targeted row crop restrictions).

Compliance costs arise on the farm level through information gathering, increased organisational tasks, and monitoring obligations on the farm[15]. These costs are assumed to be higher for incentive based policies, since farmers and agencies need to spend time on the contracting of the programmes, which is not required in the regulation options.

Additionally, detection and prosecution costs are dependent on whether a certain action (conservation tillage) or result (e.g. soil loss in tons) is awarded. Both alternatives have advantages and disadvantages (Latacz-Lohmann 2001): Voluntary conservation programmes that include the payment of subsidies have at least the advantage of showing the extent of participation through the number of contracts concluded and area covered. However, the evaluation of the environmental effect of such programmes can be rather difficult, if only agricultural measures are subsidised. This

---

[15] The technical compliance costs at the farm level are partly represented by the on-farm costs evaluated by the bio-economic modeling with MODAM. Differences in regional gross margin between status quo and conservation scenarios describe the technical compliance costs of a region. These costs represent the opportunity costs that farmers are confronted with through changes in the surrounding conditions. Technical compliance costs are not considered as administrative transaction costs and are therefore not included in these considerations.

might be the case, when the environmental effect is not strongly related to the agricultural measure. If an intensive farm extension programme is chosen, detection and prosecution costs will not occur at all. On the other hand, the effects of extension programmes are difficult to monitor, since the results of a more intense extension can only be evaluated by surveys or proxy indicators.

Table 59 assigns qualitative values to the specific forms of transaction costs for each analysed policy options. The transaction costs of all policy options analysed are expected to be similar except for some specific cost types: If programmes are spatially targeted to areas with higher erosion risks, higher enactment and implementation costs can occur, given the additional effort for the identification of erodible areas and the more specific instructions needed through the environmental agencies.

Table 59: Qualitative grading of the analysed policy options using transaction costs categories

Cost type	Incentive untargeted	Incentive targeted	Row crop restriction targeted
Enactment costs	+	++	++
Implementation costs	+++	++++	++
Compliance costs*	++	++	+
Detection costs	+	+	+
Prosecution costs	+	+	+
Overall grading	8+	10+	7+

*Compliance costs consider only the administrative on-farm costs for the incentive options
Legend: Each + represents an implicated cost module; for additional cost modules more + are added;
Source: own considerations

A targeted programme requires a more specific database at the administrating agency that contains information on where the conservation programmes should be applied. However, in terms of cost-effectiveness, a targeted approach has the advantage of focussing transaction costs on sites where soil erosion can be expected. Additionally, in the European Union, data for the administration of land use regulations (crop restriction) is already available through the IACS data, which is currently being used for the administration of both EU farm area payments and agri-environmental programmes.

The regulation approach is expected to face no costs for contracting and lower costs for policing (compliance costs), while incentives options are expected to have higher costs in these categories (see (Falconer and Whitby 1999).

Detection costs are assumed to be similar for all policy options, since all policies need monitoring facilities based on land use data. Prosecution costs are assumed to be similar, even though violations of EU agri-environmental programmes are usually sanctioned through withdrawal of payments, while land use regulations are enforced through police law.

Overall, the regulation approach shows the lowest ranking in terms of the transaction cost categories. However, since no data is available for the actual monetary amount of each category, this result provides only an orientation of the involved transaction costs and also an idea of the possibility that some cost categories might be higher for one option compared to another.

8.5.4 Attributes of participants and of the environmental good

When focussing on soil conservation, the variability of the producer costs depends on the chosen indicator for evaluating soil erosion abatement. Assuming an indicator that can show avoided soil erosion in tons per hectare, the producer costs are dependent on the technical equipment of the farm as well as on whether the soils of the farm have any potential for preventing erosion at all. High potential erosion rates mean low costs for the first ton of avoided soil erosion, if the marginal abatement costs are increasing (Meyer-Aurich and Trüggelmann 2002).

Since it is almost impossible to monitor the actual erosion value caused by a farmer as a basis for payments, soil conservation programmes are usually based on measure-oriented schemes (e.g. reduced tillage). If a certain soil conserving activity is adopted, this will be rewarded by a fixed payment.

In a regulation approach, all land users of an affected region have to comply with the regulation. Differences in their opportunity costs are only reflected in their individual losses caused by the row crop restriction.

The variability of the soil characteristics (e.g. slope, soil type, soil quality) is a central topic in this study. As Falconer (2000) had concluded, the high heterogeneity of environmental goods and the difficult definition of aims only increase the costs of conservation programmes even more.

The heterogeneity of soils in a region can be covered under spatially targeted programmes that avoid payments to applicants with no potential erosion risk at all. However, there is a trade-off between centralization and de-centralization. Increasing spatial effectiveness increases the costs of goal achievement (Urfei 1999).

In using Table 56 (p.146) as a decision support, the variability of soil qualities in the sample region justifies the use of site specific management agreements and payments. If the adoption rates of voluntary programmes are not high enough and the environmental damage is severe, more obligatory programmes would be preferable. Compensations for crop restrictions would be imaginable in such a case.

8.6 Conclusions

The results from the qualitative analysis of transaction costs types supported the regulation approach. Using the on-farm costs generated in the bio-economic model MODAM as an indicator for technical compliance costs, these were also the lowest for the regulation option of row crops, even if compensation equal to the level of on-farm cost would be assumed.

Both incentive options showed higher budget costs and on-farm costs in the modelling results (see Chapter 7) and are less favourable in terms of the qualitative ranking of the transaction costs.

Depending on the political power of the involved farmers' groups, an incentive option might be the only way to reduce the amount of erosion in an agricultural region. In such a case, it was shown that it still is more advantageous to implement a targeted option than the untargeted policy. Nevertheless, even the targeted option can be rejected for reasons of equality and fairness, i.e. all farmers should have the right to claim incentives, which would lead to the untargeted incentive scheme.

A possible way out of this dilemma is to compensate farmers within a row crop restriction scenario for some or all of the losses they face due to such restrictions. This procedure would not be considered as an incentive, but could ease the political resistance against such instruments. Again, this option brings up the question of how the compensation should be handed out spatially; should it be paid by hectare of eligible land, or based on past crop shares that had to be given up due to this policy? Even though a spatial targetedness is the common sense solution for such payments, payments to farmers are often based on historical data instead of economic considerations.

Both voluntary and regulatory approaches have advantages and disadvantages from a transaction costs perspective. As a result of this study, it should be emphasized that regulatory approaches can have advantages compared to voluntary incentive based instruments. Even though incentive instruments have been shown in some studies to be more efficient in comparison to regulatory policies, the findings of this chapter had broadened the scope of appropriate instruments for soil conservation policies. Therefore, a decision between both options should be based on the consideration of both on-farm and transaction costs.

9 Discussion

9.1 The theoretical framework

Based on the analysis of soil as a natural resource with both private and public good properties, it was concluded that society is entitled to curtail soil degradation processes in order to ensure a long-term sustainability of soil resources.

The economic framework based on Ciriacy-Wantrup's safe minimum standard (Ciriacy-Wantrup 1963) was applied for the analysis of soil as a resource, which is characterized by uncertainty in terms of its replenishment and erosion rate. Soil use that is limited to a socially agreed-on standard is more advisable than trying to find an economic optimum that might bear the risk of completely destroying the resource (Dabbert 1994), i.e. offering intergenerational equity without solving the question of how the "optimal" interest rate for the discounting of future profits is determined. The institutional aspects concerning property rights on soil use were employed as a starting point of discussion (a given set of property rights).

A cost-effectiveness analysis based on this framework was proven to be operational in the way it showed the resulting on-farm costs (opportunity costs) and budget costs of policy options. The justification of the policy options was based on the assumption that the need for conservation has already been revealed by society in the time spent on the issue in a political process, the political will to formulate relevant regulations and the provision of public money for the management and support of soil conservation programmes. The soil erosion risk values derived from a USLE model served as the selection criteria for eligible area (*spatial targeting*) for soil conservation programmes as well as decision rules for command and control solutions that manage agricultural practices.

The selection of analysed instruments (incentives, regulation) was based on the consideration of its applicability within a short time frame and a given set of property rights.

In the final chapter, the effects of transaction costs are reflected on the soil conservation programmes.

The economic framework is an appropriate tool for the analysis of implementation options of soil conservation policies based on a given set of property rights and a static analysis of the decision problem.

9.2 The bio-economic modelling approach

The combination of the soil erosion risk assessment model with a regional linear programming model provided information regarding the agricultural effects, the resulting economic implications and the soil erosion risk implied by policy changes. The model can serve as a decision support tool through the simulation of different policy conditions.

The applied soil erosion risk assessment model was effective in describing soil erosion risk originating from the natural soil conditions and the characteristics of crops and the related cropping practices, which allowed for a complex analysis based on different variables.

The USLE based assessment of soil erosion risk combined with a digital elevation model provided information on a basic grid size of 25x25 m for identifying sites that are prone to erosion in an area with short but steep slopes.

The model showed, as expected, higher erosion rates for row crops compared to winter cereals. Another crucial finding for the development of soil conservation policies is that row crops grown with soil conservation measures such as reduced tillage showed higher erosion rates than any practice of winter cereals, including those with the highest erosion risk.

The fuzzy-logic approach facilitated the transfer of the risk assessment on crops and cropping practices, which had not been tested under experimental conditions, and allowed for the quantitative comparison of such crops.

The economic regional model based on a linear programming approach reproduced the agricultural status quo sufficiently. The highly detailed description of cropping practices allowed for a precise economic and ecological assessment. The linear programming tool simulated farmers' decisions under the conditions of soil conservation policies.

The applied modelling system MODAM (Multiple Objective Decision Support Tool for Agro Ecosystem Management) was used to generate the on-farm costs of soil conservation measures on a regional level (in terms of the opportunity costs of standard production systems in the region) with an acceptable effort. The resulting budget costs (payments) were also derived from this modelling approach. The model was able to show the effects from the soil conservation policies and the general policy changes (from the conditions of Agenda 2000 to the CAP2013 reform).

The indicators chosen in this approach were helpful in the analysis of the effects of policy changes. The basic indicators *"total gross margin"* and *"soil erosion risk levels"* provided general information of the conditions resulting from the implementation of new policies. The derived indicator *"cost-effectiveness"* of soil conservation policies showed clearly which policy used the least money for the reduction of soil erosion in a region, with a distinction of where these costs occurred i.e. either as on-farm or as budget costs. The level of detail in the modelling system allowed for the specific analysis of the resulting crop shares and the spatial positioning of certain crop rotations.

However, since the model was set up as a single, regional farm, the flexibility of real farms in response to different policy changes would be overestimated. This must be taken into consideration when interpreting the results. Furthermore, behavioural aspects such as risk minimization, social influences within the community of the region and personal preferences were not taken into account. Such factors can change the adoption rate of certain practices.

Overall, the modelling approach can be applied as decision support in soil conservation policy making. It is not the intention of this study to present the model results as numbers that decisions should be based on, but more as a guideline for policy making i.e. shows possible regional effects of specific policy instruments.

9.3 The relevance of transaction costs

The inclusion of transaction costs widens the scope of the analysis for soil conservation policies. For focussing only on the budget costs of direct payments to farmers would cause the underestimation of the overall costs of a policy. Transaction costs seen as costs of (re)-defining and implementing property rights can reach considerable amounts, thus reducing the overall efficiency of a policy approach. Knowing the possible magnitude of the different types of transaction costs can help prevent costly policy choices.

However, given the methodological difficulties of measuring transaction costs and the scarcity of transaction cost data for agri-environmental policies, the relevance of transaction costs estimations is not yet at a level where well documented knowledge could be transferred and used for the detailed evaluation of policy instruments. Most studies on transaction costs of agri-environmental programmes can be seen as case studies. The results of these studies reflect more the efficiency of the involved governmental agencies than the efficiency of the programmes themselves. Furthermore, it is difficult to find an adequate indicator that the transaction costs are referred to if the compared policy instruments are diverse (e.g. incentive based programmes vs. regulations).

In the case of this study, a qualitative analysis was applied based on the general conclusions of other studies that had estimated transaction costs of agri-environmental programmes and regulations. The findings are seen as a list of arguments within the discussion of policy related transaction costs.

The results of these qualitative considerations also showed restriction policy as more advantageous than the incentive options. Nevertheless, it should be kept in mind that the cost categories were only analysed in a qualitative way. Assuming that one category is afflicted with high costs, this could outweigh the cost of all the other categories and lead to contrary results.

In general, more research is needed for the estimation of transaction costs with the focus on the generalization of the results.

9.4 Appropriate instruments for soil conservation

Both a bio-economic modelling analysis and a qualitative transaction costs reflection of three possible soil conservation policies aimed at the same level of erosion reduction within an example region in North-Eastern Germany were applied. The policies were chosen as examples based on assumptions of different sets of property rights towards the right to degrade soils. Soils were defined as a quasi non-renewable resource.

The analysed policy options were untargeted incentive on reduced tillage practices, targeted incentive on reduced tillage in areas with a higher erosion risk and targeted legal restriction on row crops in areas with high erosion risk. At this point, a general conclusion on the most efficient policy is drawn using the arguments of both bio-economic modelling and transaction costs analysis.

The modelling examples showed that all three policies can positively influence the erosion situation in a region. Given the voluntary nature of the incentive based policies, the adoption of these policies at the expected levels is subject to the attitudes of farmers.

The row crop restriction option, which bans highly erosive crops from sites with a high erosion risk, was proven to be the most effective in terms of budget and on-farm costs. However, in reality the compliance rate will depend on the threat of prosecution and expected fines for non compliance for such a policy.

The costs of policies in the modelling framework were restricted to on-farm costs and budget costs representing the sum of payments for certain measures. This restriction in the policy cost definition underestimates the costs of a legal approach such as the row crop restriction.

In reality, more costs would arise, which would consist mostly of transaction costs. It can be expected that control costs will arise in all three options, since proper compliance with restrictions and measures needs to be checked for all options. Additionally, the measures of a policy must show a clear effect on the level of erosion risk.

Whether a row crop restriction option can be implemented successfully depends on factors such as the political power of the involved stakeholder groups and the notion of equality and fairness towards affected land users. A compensation payment for a row crop restriction policy could reduce the resistance against such command and control policies. Even though spatially targeted programmes seem to be more efficient, equal treatment of farmers and equal access to payments are a political price to pay for the successful implementation of agri-environmental programmes.

From a transaction costs perspective both voluntary and regulatory approaches have advantages and disadvantages. Based on qualitative considerations, it was found that regulatory approaches were more advantageous compared to voluntary incentive based instruments, which were shown in other studies to be more efficient. However, this result could easily change if a quantitative approach, which could estimate the amount of each transaction cost category, was applied.

The most relevant criteria for a cost effective policy design are
- high effectiveness of the agricultural practice and
- close spatial correlation between programme area and erosion risk zones.

Incentive programmes related to less effective agricultural practices are very likely to show lower cost-effectiveness compared to a policy that is based on a more effective measure.

The results of this study suggest that the choice of soil conservation policies should be based both on a bio-economic modelling analysis and the reflection of the involved transaction costs for each specific case of implementation. This will provide decision makers with information on the expected costs and effects for the farms within a region and for the governmental agencies assigned with the implementation and administration of such policies.

10 Summary

The aim of the study is to analyse the economic and agricultural aspects of soil conservation and to propose instrument-measure combinations for efficient soil conservation as a decision support for the implementation of soil conservation policies. Emphasis is given to the resource and institutional economics of soil conservation.

Chapter 1 gives an introduction of this study and a description of the study region. Chapter 2 demonstrates soil functions, the definitions of soil degradation and the need for soil conservation based on the current soil conditions. Chapter 3 comprises approaches on soil conservation from the international to the national level and describes how soil conservation can be implemented.

The theoretical framework for the economic modelling approach is outlined in Chapter 4. Based on the theoretical economic analysis of soils as a natural resource, the existing property rights, the public good characteristics of soils and the resulting externalities, one is lead to the conclusion that market failure does exist. Therefore, a non-market coordination of soil use is justified. Based on the theory of a "safe minimum standard", a cost-effectiveness analysis is derived to be appropriate for the assessment of the implementation options of soil conservation policies.

Chapter 5 describes a fuzzy logic based assessment method of soil erosion risk in a sample region, which is based on an adapted USLE-approach. The approach considers both the natural conditions and the characteristics of the cropping practice. The method provides site-specific erosion risk values for standard and adjusted cropping practices, which are used as parameters in the bio-economic model.

Chapter 6 outlines the design of the applied bio-economic model MODAM. This regional linear-programming model was successfully adapted and applied to evaluate the economic and ecological effects of different policy options using the example of an agricultural region in Northeastern Germany.

Chapter 7 provides the results for a set of scenarios. The basic scenarios comprise the policy conditions of the Agenda 2000 and a CAP-reform scenario with decoupled area payments that reflect the conditions of 2013. The CAP2013-scenario serves as a comparison for the soil conservation policy scenarios. The three main scenarios on policy options include both untargeted and targeted incentives programmes for reduced tillage practices and a regulation scenario that prohibits the cultivation of highly erosive crops (row crops) on erodible soils. An optimization scenario that finds a low cost solution for different levels of soil conservation for the sample region was also calculated.

The regulation option on row crops generated almost similar cost-effective results as the optimisation option. The incentive options resulted in both high on-farm and budget costs for a similar level of erosion reduction. Based on the modelling result, the row crop restriction is the preferable policy option in terms of cost-effectiveness.

The preferability of the row crop restriction is related to another important finding of the modelling: reduced tillage practices, which are promoted by the incentive options, are less capable of reducing soil erosion risks in comparison to crop change (e.g. from row crops to cereals), which can result in a higher reduction of the erosion risk. If this result is transferred to the design of a policy, the effectiveness of a policy can increase.

Chapter 8 discusses the influence of transaction costs on the success of soil conservation programmes. The inclusion of transaction costs widens the scope of the policy analysis. Focussing only on the budget costs of direct payments to farmers would underestimate the overall costs of a policy. Transaction costs seen as costs of (re)-defining and implementing property rights can reach considerable amounts, which can reduce the overall efficiency of a policy approach. Knowing the possible magnitude of the different types of transaction costs helps prevent costly policy choices.

The results from a qualitative analysis of transaction costs also supported the row crop restriction approach. The regulation option for row crops had lower compliance costs than the incentive options. Both incentive options showed higher budget costs and on-farm costs in the modelling results and were less favourable in terms of the qualitative ranking of the transaction costs.

Chapter 9 draws some final conclusions on the theoretical framework, the bio-economic modelling approach, the relevance of transaction costs and finally, the appropriate instruments for soil conservation based on the overall results of this study.

In this study, a model was successfully developed to serve as a decision support system for the soil scientific, economic and agricultural aspects of soil conservation policies. Different policy options were compared so that the most cost-effective solution for a soil conservation policy may be found. Based on the final discussion regarding the involved transaction costs, the regulation approach was shown to be the most cost-effective option, with potentially lower transaction costs. The most relevant criteria for a cost effective policy design are high effectiveness of the agricultural practice and the spatial correlation between the programme area and the erosion risk zones. Incentive programmes related to less effective agricultural practices show lower cost-effectiveness compared to a policy that is based on a more effective measure.

Compared to other studies, the modelling approach used here is more detailed in the description of the cropping practices, which allowed for the highly specific assessment of each cropping practice. This in combination with the detailed site description (100x100 meter) provided a level of detail, which is rather high for a regional modelling approach. The inclusion of transaction costs as a final reflection of the results allowed for a broader analysis of the policy options.

11 Zusammenfassung

Ziel dieser Studie ist es, ökonomische und landwirtschaftliche Aspekte des Bodenschutzes zu analysieren sowie Instrument-Maßnahmen-Kombinationen für einen effizienten Bodenschutz als Entscheidungshilfe für die Umsetzung von Bodenschutz-Politiken vorzuschlagen. Der Schwerpunkt der Arbeit liegt dabei auf einer ressourcen- und institutionenökonomischen Betrachtung. In einem empirischen Teil wird am Beispiel einer Region in Nordostdeutschland auf der Basis von Modellrechnungen die Effizienz einzelner Politikoptionen untersucht.

Kapitel 1 enthält eine Einführung zu dieser Studie sowie eine kurze Beschreibung der ausgewählten Beispielsregion. Kapitel 2 geht auf die Funktionen des Bodens ein, liefert Definitionen der Bodendegradation und unterstreicht die Notwendigkeit des Bodenschutzes auf der Grundlage der aktuellen Bodenzustände. Kapitel 3 umfasst Konzepte für die Erhaltung der Böden auf internationaler und nationaler Ebene und beschreibt, auf welche Weise Bodenschutz implementiert werden kann.

Die theoretischen Grundlagen für einen ökonomischen Analyseansatz werden in Kapitel 4 beschrieben. Auf der Grundlage einer theoriebasierten ökonomischen Analyse werden Böden als natürliche Ressource definiert, die aufgrund der bestehenden Eigentumsrechte, den Eigenschaften von Böden als öffentlichem Gut sowie den daraus resultierenden Externalitäten den Schluss zulassen, dass ein Marktversagen bei der Steuerung einer nachhaltigen Nutzung von Böden vorliegt. Eine nicht-marktgestützte Koordinierung der Bodennutzung ist daher gerechtfertigt. Basierend auf der Theorie des "Safe Minimum Standards" wird eine Kosten-Wirksamkeits-Analyse abgeleitet, die für die Beurteilung der Umsetzung von Bodenschutzpolitiken als geeignet erscheint.

Kapitel 5 beschreibt eine Fuzzy-Logik-basierte Methode zur Bewertung des Bodenerosionsrisikos in einer Beispielsregion, die auf einem erweiterten Universal-Soil-Loss-Equation-Ansatz (USLE) basiert. Der Ansatz berücksichtigt sowohl die natürlichen Standortbedingungen als auch die Eigenschaften der landwirtschaftlichen Anbauverfahren. Eine im Vergleich zu anderen Studien sehr detaillierte Beschreibung der Anbauverfahren erlaubt eine spezifische Beurteilung der erosionsrelevanten Effekte. Dieser Ansatz in Kombination mit dem hohen Detaillierungsgrad für die Standortbeschreibung bietet eine für einen regionalen Ansatz große Genauigkeit. Die angewandte Methode generiert für das Erosionsrisiko der konventionellen und angepassten Anbauverfahren standortspezifische Werte, die als technische Parameter in ein bioökonomisches Modell eingehen.

Kapitel 6 beschreibt das Design des in dieser Studie angewendeten bioökonomischen Modells MODAM (Multi-objective decision support tool for agro-ecosystem management). Dieses regionalisierte Lineare Programmierungmodell wurde erfolgreich am Beispiel einer landwirtschaftlichen Region in Nordost-Deutschland auf die Anforderungen dieser Arbeit angepasst und für die Beurteilung der ökonomischen und ökologischen Auswirkungen unterschiedlicher Politikoptionen verwendet.

Kapitel 7 stellt die Ergebnisse einer Reihe von Szenarien dar. Die Szenarien umfassen die agrarpolitischen Rahmenbedingungen der Agenda 2000 sowie ein GAP-Reform-Szenario mit entkoppelten Zahlungen entsprechend den geplanten Bedingungen des Jahres 2013. Dieses CAP2013-Szenario dient als Referenzszenario für die Szenarien zu möglichen Bodenschutzpolitiken. Die drei Hauptszenarien zu den Politikoptionen sind 1) räumlich nicht gerichtete, 2) räumlich gerichtete Anreizprogramme für Anbauverfahren mit reduzierter Bodenbearbeitung sowie 3) ein Szenario zu einer Verordnung, die den Anbau von hoch erosiven Nutzpflanzen (Reihenkulturen) auf stärker erodierbaren Böden verbietet. Zusätzlich wurde ein Optimierungsszenario berechnet, welches nach der kostengünstigsten Lösung für eine schrittweise Anhebung der Erosionsvermeidung in der Beispielsregion sucht.

Ein Verbot von Reihenkulturen generiert ein ähnlich kosteneffizientes Ergebnis wie eine Lösung der Optimierungsoption mit vergleichbarem Erosionsniveau. Die Anreizprogramme zu pflugloser Bodenbearbeitung führen bei einer vergleichbaren Reduzierung der Bodenerosion sowohl zu höheren betrieblichen Anpassungskosten als auch zu hohen Budgetkosten. Auf der Grundlage der Modellierungsergebnisse ist ein Verbot von Reihenkulturen auf stark erodierbaren Standorten deshalb die vorzüglichere Option im Hinblick auf das Kosten-Wirksamkeits-Verhältnis.

Die Vorzüglichkeit eines Reihenkulturverbots stützt sich auf ein weiteres, wichtiges Ergebnis der Modellierung: Die Verfahren mit reduzierter Bodenbearbeitung, die durch finanzielle Anreize gefördert werden, können weniger zur Verringerung des Bodenerosionsrisikos beitragen als eine Änderung der Kulturpflanzenauswahl (z.B. von Reihenkulturen zu Getreide), welche zu einer stärkeren Reduzierung des Erosionsrisikos führt. Wenn dieses Ergebnis auf die Gestaltung der Politik übertragen wird, kann die Wirksamkeit einer Politik erhöht werden.

Kapitel 8 behandelt den Einfluss von Transaktionskosten auf den Erfolg von Bodenschutzprogrammen. Die Einbeziehung von Transaktionskosten erweitert den Betrachtungsbereich einer Politikanalyse. Wird die Betrachtung nur auf die Budgetkosten für die Direktzahlungen an die Landwirte konzentriert, werden die Gesamtkosten einer Politik unterschätzt. Transaktionskosten, verstanden als Kosten für die (Wieder-)Festlegung und Implementierung von Eigentumsrechten, können erhebliche Beträge erreichen, wodurch die Gesamteffizienz eines politischen Ansatzes reduziert werden kann. Die Kenntnis der möglichen Größenordnung der verschiedenen Arten von Transaktionskosten hilft bei der Vermeidung von kostspieligen politischen Entscheidungen.

Die Ergebnisse einer qualitativen Analyse der Transaktionskosten der untersuchten Politikoptionen unterstützen ebenfalls den Ansatz eines Verbotes von Reihenkulturen. Basierend auf qualitativen

Kostenüberlegungen bringt die Verordnungsoption (Verbot von Reihenkulturen) geringere Kosten mit sich als die beiden Anreizprogramme. Beide Anreizprogramme zeigen höhere Kosten sowohl für Budget- als auch für On-farm-Kosten (Opportunitätskosten) in den Modellierungsergebnissen und sind in Bezug auf das qualitative Ranking der Transaktionskosten mit mehr Kostenfaktoren behaftet.

Kapitel 9 zieht Schlussfolgerungen zu den theoretischen Grundlagen, dem bio-ökonomischen Modellierungsansatz, der Bedeutung der Transaktionskosten, und schließlich, zu den entsprechenden Instrumenten für die Erhaltung des Bodens, basierend auf den Gesamtergebnissen dieser Studie.

In dieser Studie wurde erfolgreich ein Modell entwickelt, welches als Entscheidungshilfe für sowohl ökonomische als auch landwirtschaftliche Aspekte des Bodenschutzes dienen kann. Unterschiedliche politische Optionen wurden im Hinblick auf eine kosteneffiziente Lösung für eine Bodenschutz-Politik untersucht. Auf der Grundlage der abschließenden Diskussion der entstehenden Transaktionskosten erweist sich der Regulierungsansatz zu Reihenkulturen als die kostengünstigste Option mit potenziell niedrigeren Transaktionskosten. Die wichtigsten Kriterien für ein kostengünstiges Politikdesign sind eine hohe Effizienz der landwirtschaftlichen Verfahren und die räumliche Korrelation zwischen dem Programmgebiet und den erosionsgefährdeten Gebieten. Anreiz-Programme im Zusammenhang mit weniger effektiven landwirtschaftlichen Praktiken weisen ein schlechteres Kosten-Nutzen-Verhältnis auf als eine Politik, die auf einer wirksameren landwirtschaftlichen Maßnahme beruht.

Der in dieser Studie zur Anwendung gekommene Modellierungsansatz weist einen im Vergleich zu anderen Arbeiten hohen Grad an Detailliertheit bei der Beschreibung der landwirtschaftlichen Anbauverfahren auf. Dieser Detailgrad erlaubt eine sehr spezifische Bewertung der Verfahren. Dies in Kombination mit der sehr genauen Standortbeschreibung (100x100 Meter) gewährleistet eine Genauigkeit der Analyse, die relativ hoch für einen regionalen Modellierungsansatz ist. Die Einbeziehung von Transaktionskosten in eine abschließende Reflektion der Ergebnisse ermöglicht eine breitere Analyse der Politikoptionen.

12 Acknowledgements

Looking back now, after finishing this thesis, I had to remember those who accompanied me on this journey, and how I finally came to writing a thesis on this subject. Many things that were unforeseeable happened. Even the subject of this thesis was not clear when I started at ZALF in 1998. My thanks are expressed in a chronological order, not by the importance of the persons involved.

All I started with in 1998 was the task to have a good look at the MODAM modeling system and to apply it in the research project GRANO that was sponsored by the BMBF. Therefore, my greatest thanks to the responsible persons of the Ministry.

I also want to thank Prof. Dr. Harald Kächele, my advisor at ZALF, for his great support and trust in putting me on this job. I will always appreciate his great humor and energy he puts into things. Without Harald I definitely would have been lost on many occasions.

At the same time, I wish to thank Prof. Dr. Klaus Mueller, the head of the Institute of Socioeconomics, for giving me the time to work on and to finish this PhD project and for willing to be my "last minute"-examiner at the final defense in 2008.

I also want to thank Dr. Peter Zander, the co-developer (together with Harald Kächele) of MODAM, for always being ready to answer questions on the basics of this modeling system.

Furthermore, I wish to thank Dr. Claudia Sattler for being such a great office-mate and for the pleasant work atmosphere.

I also want to thank everybody that was involved in my thesis at ZALF, particularly Angelika Neumeyer, Renate Wille, Kerstin Franke and Gerlinde Prentkowski for all their help and support.

In the year 2000, I finally decided to apply the MODAM approach on an economic analysis of soil conservation policies, a subject that was raised during the GRANO research project. I would like to thank all my colleagues from this project for the discussions on "better policies and measures".

During a research visit to Canada, I met Prof. PhD Alfons Weersink from the University of Guelph, Ontario. The discussions with him and his feedback on my presentations in the years following were a very valuable contribution to this study. Finally, he became the second reviewer of this thesis. I want to thank Alfons for all the great support and the warm welcome I always had whenever I visited the former AGEC (now FARE) department of the University of Guelph.

Many thanks to the friends I met during my 3 month-scholarship in Guelph in 2001, particularly Anne Hünnemeyer who introduced me to everyone in the department and a great social circle. Many thanks go to the DAAD for financing this visit.

It also took some time before I got in touch with a professor who would become my university advisor. I found this person in Prof. Dr. Stephan Dabbert from the Hohenheim University. I am very grateful for all the advice he gave me and for the support in following this idea. I also wish to thank his secretary Eva Lepper, who helped me a lot in my position as an external PhD-student.

I want to express my gratitude to Prof. Dr. Thomas Berger for being a part of the defence committee.

It was unfortunate that my mother could not be witness to the end of this study. However, I wish to thank both my mother and father for all the support they gave me and the deep trust they always had in me. Many thanks also to Ruth Ott for the caring support during the time around my defence.

Thanks to my kids Franz Chee Yin and Swee Lan Sofie for being such nice creatures, who although might have delayed things a bit, are worth it all.

Finally, I want to thank the person that had to accompany me through all the hardships of writing a thesis, my beloved "best friend and wife" Lina Peck Wan Yap. Thanks for granting me all the extra hours I needed while I was writing and for the night shifts you put in for the final proof-reading. Lina, thank you so much for your extraordinary help and understanding.

Berlin, 31.10.2008

13 References

Abelson, P. (1979): Cost benefit analysis and environmental problems. Saxon House, Farnborough.

Allen, D.W. (1991): What are transaction costs? In: R.O. Zerbe (Ed.): Research in law and economics, Elsevier, Amsterdam [u.a.], pp. 1-18.

Arriaza, M. and J.A. Gomez-Limon (2003): Comparative performance of selected mathematical programming models. Agricultural Systems 77, No. 2, pp. 155-171.

Arrow, K. (1969): The Organisation of Economic Activity: Issues Pertinent to the Choice of Market versus Non-Market Allocation. The Analysis of and Evaluation of Public expenditure 1, pp. 59-73.

Arrow, K., B. Bolin, R. Costanza, P. Dasgupta, C. Folke, C.S. Holling, B.O. Jansson, S. Levin, K.G. Maler, C. Perrings and D. Pimentel (1995): Economic-Growth, Carrying-Capacity, and the Environment. Science 268, No. 5210, pp. 520-521.

Arrow, K.J. (2001): Uncertainty and the welfare economics of medical care (reprint). Journal of Health Politics Policy and Law 26, No. 5, pp. 851-883.

Arzt, K., E. Baranek, C. Berg, K. Hagedorn, J. Lepinat, K. Müller, U. Peters, T. Schatz, C. Schleyer, R. Schmidt, J. Schuler and I. Volkmann (2002): Dezentrale Bewertungs- und Koordinationsmechanismen. In: K. Müller, V. Toussaint, H.-R. Bork, K. Hagedorn, J. Kern, U.J. Nagel, J. Peters, R. Schmidt, T. Weith, A. Werner, A. Dosch and A. Piorr (Eds.): Nachhaltigkeit und Landschaftsnutzung: neue Wege kooperativen Handelns, Margraf Verlag, Weikersheim, pp. 29-96.

Arzt, K., E. Baranek, J. Eggers, U. Fischer-Zujkow, J. Gewalter, K. Hagedorn, K. Müller, U. Peters, R. Schmidt, J. Schuler, S. Teigeler and H.-P. Weikard (2000): Projektbereich 1: Dezentrale Bewertungs- und Koordinationsmechanismen. In: K. Müller, H.-R. Bork, A. Dosch, K. Hagedorn, J. Kern, J. Peters, H.-G. Petersen, U.J. Nagel, T. Schatz, R. Schmidt, V. Toussaint, T. Weith, A. Werner and A. Wotke (Eds.): Nachhaltige Landnutzung im Konsens - Ansätze für eine dauerhaft-umweltgerechte Nutzung der Agrarlandschaften in Nordostdeutschland, Focus Verlag, pp. 21-45.

Arzt, K., E. Baranek, C. Schleyer and K. Müller (2003): Bedeutung, Modelle und Barrieren einer Regionalisierung der Agrarumweltpolitik und der Politik ländlicher Räume in der EU. Berichte über Landwirtschaft 81, pp. 208-223.

Barbier, B. and G. Bergeron (1999): Impact of policy interventions on land management in Honduras: results of a bioeconomic model. Agricultural Systems 60, No. 1, pp. 1-16.

Barbier, E.B. (1995): The Economics of Soil Erosion: Theory, Methodology and Examples. Internet, http://www.idrc.ca/uploads/user-S/10536145400ACF2B4.pdf, Last access: 21-8-2007.

Bates, R. (1989): Beyond the Miracle of the Market: The Political Economy of Agrarian Development in Kenya. Cambridge University Press, Cambridge.

Batie, S. (1989): Sustainable development: challenges to the profession of agricultural economics. American Journal of Agricultural Economics 71, No. 5, pp. 1083-1101.

Baudoux, P. (2001): Beurteilung von Agrarumweltprogrammen - eine einzelbetriebliche Analyse in Baden-Württemberg und Nordbrandenburg. Agrarwirtschaft 50, No. 4, pp. 249-261.

Bayerische Landesanstalt für Landwirtschaft (LfL) (2004): Bodenerosion - Wie stark ist die Bodenerosion auf meinen Feldern? Freising.

Bishop, R.C. (1978): Endangered species and uncertainty: the economics of a safe minimum standard. American Journal of Agricultural Economics 60, No. 1, pp. 10-18.

Bishop, R.C. (1979): Endangered Species, Irreversibility, And Uncertainty - Reply. American Journal of Agricultural Economics 61, No. 2, pp. 376-379.

BMVEL - Bundesministerium für Verbraucherschutz, E.u.L. (2005): Meilensteine der Agrarpolitik. Internet, http://www.bmelv.de/nn_751434/SharedDocs/downloads/01-Broschueren/MeilensteineAgrar,templateId=raw,property=publicationFile.pdf/MeilensteineAgrar.pdf, Last access: 22-8-2007.

Boardman, J. (1995): The impacts of erosion. In: B. Evans (Ed.): Soil erosion and land use: Towards a sustainable policy. Proceedings of the seventh Professional Environmental Seminar held on Friday 25th February 1995 at the Møller Centre, Cambridge, White Horse Press, Cambridge, pp. 3-13.

Boardman, J., J. Poesen and R. Evans (2003): Socio-economic factors in soil erosion and conservation. Environmental Science and Policy 6, No. 1, pp. 1-6.

Bork, H.-R. (1991): Bodenerosionsmodelle - Forschungsstand und Forschungsbedarf. Berichte über Landwirtschaft 205, pp. 51-57.

Bork, H.-R., C. Dalchow, H. Kächele, H.-P. Piorr and K.-O. Wenkel (1995): Agrarlandschaftswandel in Nordost-Deutschland unter veränderten Rahmenbedingungen : ökologische und ökonomische Konsequenzen. Ernst u. Sohn, Berlin.

Brand-Saßen, H. (2004): Bodenschutz in der deutschen Landwirtschaft - Stand und Verbesserungsmöglichkeiten. Dissertation, Georg-August-Universität, Göttingen.

Breton, A. (1970): Public Goods and the stability of federalism. Kyklos 23, pp. 882-902.

Bridges, E.M., I. Hannam, L.R. Oldeman, F. Penning deVries, S.J. Scherr and S. Sombatpanit (2001): Response to Land Degradation. Science Publishers Inc., Enfield, (N.H), USA.

Bromley, D.W. (1989): Economic Interests and Institutions: The Conceptual foundations of Public Policy. Blackwell, Oxford.

Buchanan, J.M., R.D. Tollison and G. Tullock (1980): Toward a Theory of the Rent-Seeking Society. College Station: Texas A & M University Press.

Bureau of Justice Assistance (2006): Center for Program Evaluation. Internet, http://www.ojp.usdoj.gov/BJA/evaluation/glossary/glossary_e.htm, Last access: 21-8-2007.

Busenkell, J. (2004): Beurteilung von Agrarumweltmaßnahmen in Nordrhein-Westfalen und Rheinland-Pfalz - Einzelbetriebliche Analyse der Programme im Ackerbau. Dissertation, Rheinische Friedrich-Wilhelms-Universität, Bonn.

Cansier, D. (1993): Umweltökonomic. Gustav Fischer, Stuttgart.

Challen, R. (2000): Institutions, Transaction Costs and Environmental Policy. Cheltenham, Northhampton.

Cicchetti, C. and L. Wilde (1992): Uniqueness, irreversability, and the theory of nonuse values. American Journal of Agricultural Economics 74, No. 5, pp. 1121-1125.

Ciriacy-Wantrup, S.V. (1963): Resource conservation : economics and policies. rev. ed. Univ. of Calif. Press, Berkeley.

Ciriacy-Wantrup, S.V. and R. Bishop (1975): "Common property" as a concept in natural resources policy. Natural Resources Journal 15, pp. 713-727.

Clark, E.H. (1985): The Off-site Costs of Soil Erosion. Journal of Soil and Water Conservation 40, No. 1, pp. 19-22.

Clark, R. (1996): Methodologies for the economic analysis of soil erosion and conservation. CSERGE Working Paper No. GEC 96-13.

Coase, R. (1937): The Nature of the Firm. Economica (New Series) 4, No. 16, pp. 386-405.

Coase, R. (1960): The Problem of Social Cost. Journal of Law and Economics 3, pp. 1-44.

Colombo, S., J. Calatrava-Requena and N. Hanley (2003): The economic benefits of soil erosion control: An application of the contingent valuation method in the Alto Genil basin of southern Spain. Journal of Soil and Water Conservation 58, No. 6, pp. 367-371.

Commission of the European Communities (2002): Towards a Thematic Strategy for Soil Protection. Communication from the Commission to the Council, the European Parliament, the Economic and Social Committee and the Committee of the Regions. Brussels. COM (2002) 179 final.

Commission of the European Communities (2006): Thematic Strategy for Soil Protection - COM(2006)231 final. Internet, http://ec.europa.eu/environment/soil/pdf/com_2006_0231_en.pdf, Last access: 21-8-2007.

Crosson, P. (1984): New perspective on soil conservation policy. Journal of Soil and Water Conservation 39(4), pp. 222-225.

Crowards, T.M. (1998): Safe Minimum Standards: costs and opportunities. Ecological Economics 25, No. 3, pp. 303-314.

Cummings, R., P. Ganderton and T. McGuckin (1994): Substitution effects in CVM values. American Journal of Agricultural Economics 76, No. 2.

Dabbert, S. (1994): Ökonomik der Bodenfruchtbarkeit. Stuttgart.

Dabbert, S., S. Herrmann, G. Kaule and M. Sommer (1999): Landschaftsmodellierung für die Umweltplanung : Methodik, Anwendung und Übertragbarkeit am Beispiel von Agrarlandschaften. Springer-Verlag, Berlin [u.a.].

Dantzig, G.B. (1963): Linear Programming and Extensions. Princeton University Press.

Der Rat von Sachverständigen für Umweltfragen (SRU) (1996): Konzepte einer dauerhaft-umweltgerechten Nutzung ländlicher Räume. Metzler-Poeschel, Stuttgart.

Deumlich, D., J. Thiere and L. Voelker (1996): Assessing the potential water erosion risk of natural areas, administrative regions and communities in NE Germany. Mitteilungen der Deutschen Bodenkundlichen Gesellschaft 79.

Deumlich, D., J. Thiere and L. Völker (1997): Vergleich zweier Methoden zur Beurteilung der Wassererosionsgefährdung von Wassereinzugsgebieten. Wasser und Boden 49, No. 5, pp. 46-51.

Deybe, D. and G. Flichman (1991): A regional agricultural model using a plant-growth simulation program as activities generator - an application to a region in Argentina. Agricultural Systems 37, No. 4, pp. 369-385.

DirektZahlVerpflV (2006): Verordnung über die Grundsätze der Erhaltung landwirtschaftlicher Flächen in einem guten landwirtschaftlichen und ökologischen Zustand (Direktzahlungen-Verpflichtungenverordnung; Bundesgesetzblatt Jahrgang 2004 Teil I Nr. 58, ausgegeben zu Bonn am 12. November 2004. Internet, http://www.gesetze-im-internet.de/bundesrecht/direktzahlverpflv/gesamt.pdf, Last access: 21-8-2007.

DLG (1991): DLG-Futterwerttabellen : Schweine. DLG-Verlag, Frankfurt [M].

DLG (1993): Faustzahlen für Landwirtschaft und Gartenbau. Bugra Suisse, München, Bern.

DLG-Futterwerttabellen Wiederkäuer (1997): DLG-Verlag, Frankfurt [M].

Chapter 13 - References

Donaldson, A.B., G. Flichman and J.P.G. Webster (1995): Integrating agronomic-models and economic-models for policy analysis at the farm-level - the impact of cap reform in 2 European regions. Agricultural Systems 48, No. 2, pp. 163-178.

Drake, L., P. Bergström and H. Svedsäter (1999): Farmers' attitudes and uptake. In: G.v. Huylenbroeck and M. Whitby (Eds.): Countryside Stewardship: Farmers, Policies and Markets, Amsterdam, pp. 89-111.

DüV (2006): Düngeverordnung - Verordnung über die Anwendung von Düngemitteln, Bodenhilfsstoffen, Kultursubstraten und Pflanzenhilfsmitteln nach den Grundsätzen der guten fachlichen Praxis beim Düngen - In der Fassung der Bekanntmachung vom 10. Januar 2006, BGBl.(Bundesgesetzblatt). Internet, Last access: 21-8-2007.

Edwards, S.R. and Anderson.G.D. (1987): Overlooked Biases in Contingent Valuation Surveys: Some Considerations. Land Economics 62, No. 2, pp. 168-178.

Enquete-Kommission (1994): Schutz der Grünen Erde. Klimaschutz durch umweltgerechte Landwirtschaft und Erhalt der Wälder. Economica Verlag, Bonn.

EPA - US Environmental Protection Agency (2006): Clean Air Markets - Programs and Regulations. Internet, http://www.epa.gov/airmarkets/progsregs/index.html, Last access: 21-8-2007.

Ervin, D.E. and J.W. Mill (1985): Agricultural Land Markets And Soil-Erosion - Policy Relevance And Conceptual Issues. American Journal of Agricultural Economics 67, No. 5, pp. 938-942.

European Commission (2006): Agenda 2000. Internet, http://ec.europa.eu/agenda2000/overview/en/agenda.htm, Last access: 21-8-2007.

European Council (2003): Council Directive of 27 June 1985 on the assessment of the effects of certain public and private projects on the environment 85/337/EEC. Internet, http://europa.eu.int/comm/environment/eia/full-legal-text/85337.htm, Last access: 22-8-2007.

European Council (2006a): Commission Regulation (EC) No 796/2004 of 21 April 2004 (consolidated version) laying down detailed rules for the implementation of cross-compliance, modulation and the integrated administration and control system provided for in of Council Regulation (EC) No 1782/2003 establishing common rules for direct support schemes under the common agricultural policy and establishing certain support schemes for farmers. Internet, http://eurlex.europa.eu/LexUriServ/LexUriServ.do?uri=CONSLEG:2004R0796:20050325:EN:PDF, Last access: 21-8-2007a.

European Council (2006b): COUNCIL REGULATION (EC) No 1257/1999 of 17 May 1999 on support for rural development from the European Agricultural Guidance and Guarantee Fund (EAGGF) and amending and repealing certain Regulations (OJ L 160, 26.6.1999, p. 80). Internet, http://eur-lex.europa.eu/LexUriServ/LexUriServ.do?uri=CONSLEG:1999R1257:20040501:EN:PDF, Last access: 21-8-2007b.

European Council (2006c): Council Regulation (EC) No 1782/2003 of 29 September 2003 (consolidated version) establishing common rules for direct support schemes under the common agricultural policy and establishing certain support schemes for farmers and amending Regulations (EEC) No 2019/93, (EC) No 1452/2001, (EC) No 1453/2001, (EC) No 1454/2001, (EC) 1868/94, (EC) No 1251/1999, (EC) No 1254/1999. Internet, http://eurlex.europa.eu/LexUriServ/LexUriServ.do?uri=CONSLEG:2003R1782:20060101:EN:PDF, Last access: 21-8-2007c.

European Council (2008a): Council Regulation (EC) No 1698/2005 of 20 September 2005 on support for rural development by the European Agricultural Fund for Rural Development (EAFRD). Internet, http://europa.eu.int/eurlex/lex/LexUriServ/site/en/oj/2005/l_277/l_27720051021en00010040.pdf, Last access: 13-3-2008a.

European Council (2008b): Press release PRES/07/286; 2842nd Council meeting; Brussels, 20 December 2007. Internet, http://europa.eu/rapid/pressReleasesAction.do?reference=PRES/07/286&format=PDF&aged=0&language=EN&guiLanguage=en, Last access: 28-2-2008b.

Evans, R. (1995): Soil erosion and land use: Towards a sustainable policy. In: B. Evans (Ed.): Soil erosion and land use: Towards a sustainable Policy. Proceedings of the seventh Professional Environmental Seminar held on Friday 25th February 1995 at the Møller Centre, Cambridge, White Horse Press, Cambridge, pp. 14-26.

Evans, R. (1996): Soil erosion and its impacts in England and Wales. Friends of the Earth Trust, London.

Falconer, K. (2000): Farm-level constraints on agri-environmental scheme participation: a transactional perspective. Journal of Rural Studies v 16 (3), pp. 379-394.

Falconer, K., P. Dupraz and M. Whitby (2001): An investigation of policy administrative costs using panel data for the English Environmentally Sensitive Areas. Journal of Agricultural Economics Vol.52, No.1, pp. 83-103.

Falconer, K. and C. Saunders (2002): Transaction costs for SSSIs and policy design. Land Use Policy 19, No. 2, pp. 157-166.

Falconer, K. and M. Whitby (1999): The invisible costs of scheme implementation and administration. In: G.v. Huylenbroeck and M. Whitby (Eds.): Countryside Stewardship: Farmers, Policies and Markets, Amsterdam, pp. 67-88.

Falconer, K. and M. Whitby (2000): Untangling red tape: scheme administration and the invisible costs of European agri-environmental policy. European Environment 10, No. 4, pp. 193-203.

Federal Ministry for the Environment, N.C.a.N.S. (2003a): Federal Soil Protection Act. Internet, http://www.bmu.de/files/soilprotectionact.pdf, Last access: 11-7-2003a.

Federal Ministry for the Environment, N.C.a.N.S. (2003b): Federal Soil Protection and Contaminated Sites Ordinance. Internet, http://www.bmu.de/english/download/soil/files/bbodschv_uk.pdf, Last access:

Forster, D.L. (2000): Public policies and private decisions: Their impacts on Lake Erie water quality and farm economy. Journal of Soil and Water Conservation 55, No. 3, pp. 309-322.

Forster, D.L. and J.N. Rausch (2002): Evaluating agricultural nonpoint-source pollution programs, in two Lake Erie tributaries. Journal of Environmental Quality 31, No. 1, pp. 24-31.

Fox, G., G. Umali and T. Dickinson (1995): An Economic Analysis of Targeting Soil Conservation Measures with Respect to Off-site Water Quality. Canadian Journal of Agricultural Economics/Revue canadienne d'agroeconomie 43, No. 1, pp. 105-118.

Frede, H.-G. and S. Dabbert (1998): Handbuch zum Gewässerschutz in der Landwirtschaft. ecomed, Landsberg.

Freshfields (2003): Environmental liability in Germany. Freshfields Bruckhaus Deringer, Internet, http://freshfields.com/practice/environment/publications/pdfs/2858.pdf, Last access: 25-10-2003.

Frielinghaus, M., A. Kocmit, H.-R. Bork and R. Schmidt (1998): Tolerierbarer Bodenabtrag - Grenzen seiner Anwendbarkeit. Mitteilungen der Deutschen Bodenkundlichen Gesellschaft 88, pp. 565-567.

Frielinghaus, M., D. Deumlich, R. Funk, K. Helming, H. Kretschmer, R. Roth, J. Thiere, L. Völker and B. Winnige (1997): Merkblätter zur Bodenerosion in Brandenburg. ZALF-Bericht Nr. 27.

Furtan, W.H. and S.S. Hosseini (2003): Economic and Institutional Considerations for Soil Depletion. Internet, http://www.csale.usask.ca/PDFDocuments/considerSoilDepletion.pdf, Last access: 21-8-2007.

Grepperud, S. (2000): Optimal soil depletion with output and price uncertainty. Environment and Development Economics 5, No. 3, pp. 221-240.

Griffin, R.C. and D.W. Bromley (1982): Agricultural runoff as a nonpoint externality: a theoretical development. American Journal of Agricultural Economics 64, No. 3, pp. 547-552.

Grimm, M., R. Jones and L. Montanarella (2002): Soil Erosion Risk in Europe. European Soil Bureau, Institute for Environment & Sustainability, JRC, Ispra.

Hagedorn, K., V. Beckmann, S. Tiemann, and K. Reuter (2004). Kosten der Erreichung von Umweltqualitätsziele in ausgewählten Regionen durch Umstellung auf Ökologischen Landbau im Vergleich zu anderen Agrarumweltmaßnahmen unter besonderer Berücksichtigung von Administrations- und Kontrollkosten, Bericht 02OE227, Bundesprogramm Ökologischer Landbau.

Hampicke, U. (1991): Naturschutz-Ökonomie. Ulmer, Stuttgart.

Hannam, I. and B. Boer (2002): Legal and Institutional Frameworks for Sustainable Soils. UICN Environmental Policy Paper No. 45. Internet, http://www.iucn.org/themes/law/pdfdocuments/EPLP45EN.pdf, Last access: 21-8-2007.

Hansen, L.T., V.E. Breneman, C.W. Davison and C.W. Dicken (2002): The cost of soil erosion to downstream navigation. Journal of Soil and Water Conservation 57, No. 4, pp. 205-212.

Hanusch, H. (1987): Nutzen-Kosten-Analyse. Franz Vahlen, München.

Hartmann, E., A. Schekahn, R. Luick and F. Thomas (2006): Kurzfassungen der Agrarumwelt- und Naturschutzprogramme - Darstellung und Analyse von Maßnahmen der Agrarumwelt- und Naturschutzprogramme in der Bundesrepublik Deutschland. Internet, Last access: 21-8-2007.

Heckelei, T. (2002): Calibration and Estimation of Programming Models for Agricultural Supply Analysis - Habilitation Thesis. University of Bonn, Germany.

Hediger, W. (2003): Sustainable farm income in the presence of soil erosion: an agricultural Hartwick rule. Ecological Economics 45, No. 2, pp. 221-236.

Held, M. and K. Kümmerer (1997): Preserving soils for life. Gaia.

Henrichsmeyer, W., O. Gans and I. Evers (1991): Einführung in die Volkswirtschaftslehre. Ulmer, Stuttgart.

Houghton, P.D. and P.E.V. Charman (1986): Glossary of Terms Used in Soil Conservation. Soil Conservation Service of New South Wales and the Standing Committee on Soil Conservation.

Howitt, R.E. (1995): Positive mathematical programming. American Journal of Agricultural Economics 77 (1995) 2, pp. 329-342.

Hubbard, M. (1997): The 'New Institutional Economics' in Agricultural Development: Insights and Challenges. Journal of Agricultural Economics 48 (2), pp. 239-249.

Hurni, H. (2003): Current international actions for furthering the sustainable use of soils; symposium no.61 paper no. 1855; 17th WCSS (World Congress of Soil Science), 14-21 August 2002, Bangkok, Thailand. Internet, http://www.ldd.go.th/Wcss2002/papers/1855.pdf, Last access: 14-8-2002.

Huylenbroeck, G.v. and M. Whitby (1999): Countryside stewardship: farmers, policies, and markets. Pergamon, Amsterdam [u.a.].

ISRIC - World Soil Information (2006): Global Assessment of Land Degradation and Improvement (GLADA). Internet, http://www.isric.org/UK/About+ISRIC/Projects/Current+Projects/GLADA.htm, Last access: 20-8-2007.

Jarosch, J. and J. Zeddies (1991): Bodenerosion: Ökonomische Aspekte von Schaden und Schutzmassnahmen - Soil erosion: economic aspects of damage and protective measures. Berichte uber Landwirtschaft, Sonderheft 205, No. 3, pp. 99-116.

Kächele, H. (1999): Auswirkungen großflächiger Naturschutzprojekte auf die Landwirtschaft. Ökonomische Bewertung der einzelbetrieblichen Konsequenzen am Beispiel des Nationalparks "Unteres Odertal". Frankfurt.

KASSA (2006): KASSA - Knowledge Assessment and Sharing on Sustainable Agriculture - The European Platform. Internet, http://kassa.cirad.fr/content/download/847/3245/file/KASSA_BEST_OF_EUR_02.pdf, Last access: 20-8-2007.

Khan, R. (1993): International Law of Land Degradation. International Studies 30, No. 3, pp. 255-275.

Kiker, C. and G. Lynne (1986): An Economic-Model of Soil Conservation - Comment. American Journal of Agricultural Economics 68, No. 3, pp. 739-742.

Knight, J. (1992): Institutions and Social Conflict. Cambridge University Press.

Kooten, G.C.v. (1993): Land resource economics and sustainable development - economic policies and the common good. UBC Press, Vancouver.

Kraemer, R.A., R. Landgrebe-Trinkunaite, T. Dräger, B. Görlach, N. Kranz and M. Verbücheln (2006): EU Soil Protection Policy: Current Status and the Way Forward. Ecologic - Institute for International and European Environmental Policy, Internet, http://www.ecologic.de/download/projekte/1950-1999/1965/1965_background_paper.pdf, Last access: 21-8-2007.

Krayl, E. (1993): Strategien zur Verminderung der Stickstoffverluste aus der Landwirtschaft. Wissenschaftsverlag Vauk, Kiel.

KTBL (1998): KTBL-Taschenbuch Landwirtschaft : Daten für die Betriebskalkulation in der Landwirtschaft. Landwirtschaftsverlag, Münster-Hiltrup.

KTBL (2000): Datensammlung für die Betriebsplanung in der Landwirtschaft. Landwirtschaftsverlag, Münster.

Kula, E. (1992): Economics of natural resources and the environment. Chapman & Hall, London; New York.

Laflen, J.M., L.J. Lane and G.R. Foster (1991): WEPP. A new technology of erosion prediction technology. Journal of Soil and Water Conservation 46, No. 1, pp. 34-38.

Landel, C., R. Vogg and C. Wüterich (1998): Bundes-Bodenschutzgesetz : Textausgabe mit Einführung und Lexikon zum neuen Bodenrecht. Müller Verlag, Heidelberg.

Landesamt für Verbraucherschutz und Landwirtschaft Brandenburg (2003a): ISLARA - Viehbesatz. Internet, http://www.brandenburg.de/lelf/a2/d24/islara/extra/vieh100ha.htm, Last access: 18-10-2003.

Landesamt für Verbraucherschutz und Landwirtschaft Brandenburg (2003b): Landkreis Uckermark; Milchleistung kg/Kuh und Jahr. Internet, http://www.brandenburg.de/lelf/a2/d24/islara/html/f6r.htm, Last access: 18-10-2003.

Landesbetrieb für Datenverarbeitung und Statistik (2006): Statistische Berichte A V 3 - 4j / 04 Flächenerhebung nach Art der tatsächlichen Nutzung im Land Brandenburg 2004. Internet, http://www.mdf.brandenburg.de/sixcms/media.php/4055/AV3_4j-04_ebook.pdf, Last access: 20-8-2007.

Landesumweltamt Brandenburg (LUA) (2002): A.N.R.N.A.u.B. Biotopkartierung Brandenburg. 2002.

Landkreis Uckermark (2006): Jahresbericht 2004 - Zahlen und Fakten zur Arbeit des Landwirtschafts- und Umweltamtes -. Internet, Last access: 20-8-2007.

Lankoski, J. and M. Ollikainen (2003): Agri-environmental externalities: a framework for designing targeted policies. European Review of Agricultural Economics 30, No. 1, pp. 51-75.

Latacz-Lohmann, U. (2001): A policy decision-making framework for devising optimal implementation strategies for good agricultural and environmental policy practices. OECD Papers 1, No. 1, pp. 1-33.

Latacz-Lohmann, U. (2004): Dealing with limited information in designing and evaluating agrienvironmental policy. Internet, http://merlin.lusignan.inra.fr:8080/eaae/website/pdf/121_Latacz, Last access: 20-8-2007.

Lazo, J.K., W.D. Schulze, G.H. Mcclelland and J.K. Doyle (1992): Can Contingent Valuation Measure Nonuse Values? American Journal of Agricultural Economics 74, No. 5, pp. 1126-1132.

Levin, H.M. and P.J. McEwan (2001): Cost-effectiveness analysis. Sage Publications, Inc., ThousandOaks.

Lintner, A. and A. Weersink (1999): Endogenous Transport Coefficients: Implications for Improving Water Quality from Multi-Contaminants in an Agricultural Watershed. Environmental and Resource Economics 14, No. 2, pp. 269-296.

Lippert, C. (1999): Institutionenökonomische Überlegungen zur optimalen Bereitstellung und Entlohnung von Umweltattributen in Agrarlandschaften - Optimal Provision and Remuneration for Environmental Attributes of Agricultural Landscapes seen from an Institutional Economics Point of View. Agrarwirtschaft 48 (1999) 11, pp. 417-430.

Lippert, C. (2005): Institutionenökonomische Analyse von Umwelt- und Qualitätsproblemen des Agrar- und Ernährungssektors. Wiss.-Verl. Vauk, Kiel.

Lu, Y. and M. Stocking (1998): A decision-support model for soil conservation: case study on the Loess Plateau, China. CSERGE Working Paper No. No. WM 98-04.

Margolis, M. and E. Nævdal (2004): Safe Minimum Standards in Dynamic Resource Problems-Conditions for Living on the Edge of Risk. Ressources for the Future, Internet, http://www.rff.org/rff/Documents/RFF-DP-04-03.pdf, Last access: 20-8-2007.

Markandya, A. and D.W. Pearce (1991): Development, the Environment, and the Social Rate of Discount. The World Bank Research Observer 6, No. 2, pp. 137-152.

Masutti, C. (2004): Le Dust Bowl, la politique de conservation des ressources et les écologues aux Etats-Unis dans les années 1930. Strasbourg.

Matthews, R.C.O. (1986): The economics of institutions and the sources of growth. Economic Journal 96, pp. 903-918.

Matzdorf, B. and A. Piorr (2003): Beschreibung des Programms + Anhang. In: ZALF Müncheberg (Projektleitung) (Ed.): Halbzeitbewertung des Plans zur Entwicklung des ländlichen Raums des

Landes Brandenburg (im Auftrag des Ministeriums für Landwirtschaft, Umweltschutz und Raumordnung des Landes Brandenburg), pp. 9-34.

Matzdorf, B., A. Piorr and C. Sattler (2003): Halbzeitbewertung des Plans zur Entwicklung des ländlichen Raums (EPLR) gemäß VO (EG) Nr. 1257/1999 des Landes Brandenburg. Leibniz-Zentrum für Agrarlandschafts- und Landnutzungsforschung, Müncheberg.

McCann, L., B. Colby, K.W. Easter, A. Kasterine and K.V. Kuperan (2005): Transaction cost measurement for evaluating environmental policies. Ecological Economics 52, No. 4, pp. 527-542.

McCann, L. and K.W. Easter (1998): Estimating Transaction Costs of Alternative Policies to Reduce Phosphorous Pollution in the Minnesota River - Staff Paper P98-7. Department of Applied Economics, Internet, http://agecon.lib.umn.edu/cgi-bin/pdf_view.pl?paperid=935&ftype=.pdf, Last access: 20-8-2007.

McCann, L. and K.W. Easter (1999a): Differences Between Farmer And Agency Attitudes Regarding Policies To Reduce Phosphorus Pollution In The Minnesota River Basin. Review of Agricultural Economics 21, No. 1, pp. 189-207.

McCann, L. and K.W. Easter (1999b): Transaction costs of policies to reduce agricultural phosphorus pollution in the Minnesota River. Land Economics 75 No. 3, pp. 402-414.

McCann, L. and K.W. Easter (2000): Estimates of Public Sector Transaction Costs in NRCS Programs. Journal of Agricultural & Applied Economics 32, No. 3, pp. 555-563.

McCann, L. and K.W. Easter (2004): A framework for estimating the transaction costs of alternative mechanisms for water exchange and allocation. Water Resources Research 40, No. 9, pp. 1-6.

McConnell, K.E. (1983): An economic model of soil conservation. American Journal of Agricultural Economics 65, pp. 83-89.

Meyer-Aurich, A. (2001): Entwicklung von umwelt- und naturschutzgerechten Verfahren der landwirtschaftlichen Landnutzung für das Biosphärenreservat Schorfheide-Chorin. Verlag Agrarökonomie, Bern, Hannover.

Meyer-Aurich, A. (2005): Economic and environmental analysis of sustainable farming practices - a Bavarian case study. Agricultural Systems 86, No. 2, pp. 190-206.

Meyer-Aurich, A. and L. Trüggelmann (2002): Finding the optimal balance between economical and ecological demands on agriculture - research results and model calculations for a Bavarian experimental farm. Technische Universität München, Internet, http://www.weihenstephan.de/%7Eameyer/papers/AARES_paper.PDF, Last access: 20-8-2007.

Meyer-Aurich, A., P. Zander, A. Werner and R. Roth (1998): Developing agricultural land use strategies appropriate to nature conservation goals and environmental protection. Landscape and Urban Planning 41, No. 2, pp. 119-127.

Ministerium für Ländliche Entwicklung, U.u.V.d.L.B.M. (2006): Agrarbericht 2005 zur Land- und Ernährungswirtschaft des Landes Brandenburg. Internet, http://www.mluv.brandenburg.de/cms/media.php/2320/agb_2005.pdf, Last access: 20-8-2007.

Musgrave, R.A. and P.B. Musgrave (1989): Public Finance in Theory and Practice. McGraw-Hill, New York, London.

Nakao, M. and B. Sohngen (2000): The effect of site quality on the costs of reducing soil erosion with Riparian buffers. Journal of Soil and Water Conservation 55, No. 2, pp. 231-237.

Navrud, S. (2000): Valuation Techniques and Benefit Transfer Methods: Strengths, Weaknesses and Policy Utility. Valuing Rural Amenities, Organisation for Economic Co-operation and Development (OECD), Paris, pp. 15-34.

North, D. (1990): Institutions, Institutional Change and Economic Performance. Cambridge University Press, Cambridge.

Oates, W.E. and P.R. Portney (2001): The Political Economy of Environmental Policy. Resources for the Future, Internet, http://www.rff.org/Documents/RFF-DP-01-55.pdf, Last access: 20-8-2007.

OECD (2001): Environmental Indicators for Agriculture. Paris.

Oldeman, L.R., V.W.P. van Engelen and J.H.M. Pulles (1990): The extent of human-induced soil degradation. In: L.R. Oldeman, R.T.A. Hakkeling and W.G. Sombroek (Eds.): World Map of the Status of Human – Induced Soil Degradation, an exploratory note, Wageningen.

Oppermann, R. and H.U. Gujer (2003): Artenreiches Grünland bewerten und fördern. Ulmer, Stuttgart.

Ostrom, E. (1991): Governing the Commons: the evolution of institutions for collective action. Cambridge University Press.

Paris, Q. (1991): An economic interpretation of linear programming. Iowa State University Press.

Paris, Q. and R.E. Howitt (1998): An analysis of ill-posed production problems using maximum entropy. American Journal of Agricultural Economics Volume 80, Number 1, pp. 124-138.

PAV (2006): Pflanzenschutz-Anwendungsverordnung in der Fassung der Verordnung zur Bereinigung pflanzenschutzrechtlicher Vorschriften vom 10. November 1992 (BGBl. I S. 1887). Internet, http://bundesrecht.juris.de/pflschanwv_1992/BJNR118870992.html, Last access: 20-8-2007.

Pearce, D.W. (1993): Economic values and the natural world. Earthscan Publications, London.

Pearce, D.W., B. Groom, C. Hepburn and P. Koundouri (2003): Valuing the Future. World Economics 4, No. 2, pp. 121-141.

Pearce, D.W., A. Markandya and E.B. Barbier (1990): Blueprint for a green economy. London.

Pearce, D.W. and R.K. Turner (1990): Economics of natural resources and the environment. New York.

Pearce, D.W. and J.J. Warford (1993): World without end: economics, environment, and sustainable development. Oxford University Press, New York, N.Y.

Pezzey, J.C.V. and M.A. Toman (2002): The Economics of Sustainability: A Review of Journal Articles. Resources for the Future, Internet, http://www.rff.org/Documents/RFF-DP-02-03.pdf, Last access: 20-8-2007.

Pigou, A.C. (1998): Wealth and welfare - Faksimile der 1912 in London erschienenen Erstausgabe. Verl. Wirtschaft und Finanzen, Düsseldorf.

Pimentel, D., C. Harvey, P. Resodudarmo, K. Sinclair, D. Kurz, M. McNair, S. Crist, L. Shpritz, L. Fitton, R. Saffouri and R. Blair (1995): Environmental and economic costs of soil erosion and conservation benefits. Science (Washington) 267, No. 5201.

Plankl, R. (1999): Honorierung ökologischer Leistungen - Erfahrungen mit dem US-amerikanischen "Conservation Reserve Program" (CRP). Agrarstruktur und ländliche Räume: Rückblick und Ausblick; Festschrift zum 65. Geburtstag von Eckhart Neander, pp. 163-175.

Popp, J., D. Hoag and J. Ascough (2002): Targeting soil-conservation policies for sustainability: new empirical evidence. Journal of Soil and Water Conservation Vol.57, No.2, pp. 66-74.

Pretty, J.N., C. Brett, D. Gee, R.E. Hine, C.F. Mason, J.I.L. Morison, H. Raven, M.D. Rayment and G.v.d. Bijl (2000): An assessment of the total external costs of UK agriculture. Agricultural Systems 65, No. 2, pp. 113-136.

Regionomica (2006): Abschlussbericht Wirtschaftsrahmenplan Uckermark. Internet, http://www.uckermark.de/media/custom/553_1332_1.PDF, Last access: 21-8-2007.

Renard, K.G., G.R. Foster, G.A. Weesies and J.P. Porter (1991): RUSLE. Revised Universal Soil Loss Equation. Journal of Soil and Water Conservation 46, No. 1, pp. 30-33.

Roedenbeck, I. (2004): Bewertungskonzepte für eine nachhaltige und umweltverträgliche Landwirtschaft; Fünf Verfahren im Vergleich. FG Landwirtschaft, Hamburg.

Röhm, O. (2001): Analyse der Produktions- und Einkommenseffekte von Agrarumweltprogrammen unter Verwendung einer weiterentwickelten Form der positiven quadratischen Programmierung. Shaker, Aachen.

Röhm, O. and S. Dabbert (2003): Integrating agri-environmental programs into regional production models: An extension of positive mathematical programming. American Journal of Agricultural Economics 85, No. 1, pp. 254-265.

Rudloff, B. and G. Urfei (2000): Agrarumweltpolitik nach dem Subsidiaritätsprinzip. Bd. 3: Kategorisierung von Umwelteffekten und Evaluierung geltender Politikmaßnahmen. Analytica, Berlin.

Rürup, B., W. Sesselmeier and M. Enke (2002): Fischer-Wirtschaftslexikon. Fischer-Taschenbuch-Verlag, Frankfurt am Main.

Sattler, C. (2007): Evaluation of ecological effects of cropping activities; Dissertation in print. Humboldt-Universität Berlin.

Schachtschabel, P., H.-P. Blume, G. Brümmer, K.-H. Hartge and U. Schwertmann (1992): Lehrbuch der Bodenkunde - Scheffer/Schachtschabel. Ferdinand Enke Verlag, Stuttgart.

Scheele, M., F. Isermeyer and G. Schmitt (1992): Umweltpolitische Strategien zur Lösung der Stickstoffproblematik in der Landwirtschaft. Institut für Betriebswirtschaft der Bundesforschungsanstalt für Landwirtschaft, Braunschweig-Völkenrode.

Schmidt, R. and R. Diemann (1981): Erläuterungen zur Mittelmaßstäbigen Landwirtschaftlichen Standortkartierung (MMK). FZ f. Bodenfruchtbarkeit, Ber. Eberswalde, Müncheberg.

Schmidt, R. and H. Müller (1992): Einführung in das Umweltrecht. Beck, München.

Schuler, J. and H. Kächele (2003): Modelling on-farm costs of soil conservation policies with MODAM. Environmental Science and Policy 6, No. 1, pp. 51-55.

Schwertmann, U., W. Vogl and M. Kainz (1987): Bodenerosion durch Wasser - Vorhersage des Abtrags und Bewertung von Gegenmaßnahmen. Verlag Eugen Ulmer, Stuttgart.

Scott, A.D. (1989a): Conceptual origins of the rights of fishing. Kluwer Academic Publishers, Dordrecht.

Scott, A.D. (1989b): Evolution of individual transferable quotas as a distinct class of property right. In: H. Campbell, K. Menz and G. Waugh (Eds.): Economics of Fishery Management in the Pacific Islands Region, ACIAR Proceedings, pp. 51-67.

Segerson, K. (1988): Uncertainty and incentives for nonpoint pollution control. Journal of Environmental Economics and Management 15, No. 1, pp. 87-98.

Segerson, K. and T.J. Miceli (1998): Voluntary Environmental Agreements: Good or Bad News for Environmental Protection? Journal of Environmental Economics and Management 36, pp. 109-130.

Shankar, B., E.A. DeVuyst, D.C. White, J.B. Braden and R.H. Hornbaker (2000): Nitrate abatement practices, farm profits, and lake water quality: a central Illinois case study. Journal of Soil and Water Conservation 55, No. 3, pp. 296-303.

Shortle, J.S. and J.A. Miranowski (1987): Intertemporal Soil Resource Use: Is it Socially Excessive? Journal of Environmental Economics and Management 14, pp. 99-111.

Smith, A. (2005): Untersuchung über Wesen und Ursachen des Reichtums der Völker - An inquiry into the nature and causes of the wealth of nations. UTB, Tübingen.

Statistisches Bundesamt Deutschland (2006): Löhne und Gehälter - Durchschnittliche Bruttoverdienste in den neuen Ländern und Berlin-Ost. Internet, http://www.destatis.de/jetspeed/portal/cms/Sites/destatis/Internet/DE/Content/Statistiken/Verdiens teArbeitskosten/Bruttoverdienste/Tabellen/Content50/LandwirtschaftNL.psml, Last access: 20-8-2007.

Stiglitz, J. (1974): Incentives and Risk-Sharing in Sharecropping. Review of Economic Studies 41, pp. 219-255.

Stiglitz, J. (1986): The New Development Economics. World Development 14 (2), pp. 257-265.

Stone, R. (2000): Factsheet - ISSN 1198-712X Universal Soil Loss Equation (USLE). http://www.omafra.gov.on.ca/english/engineer/facts/00-001.htm#tab6, Last access: 28-8-2006.

Storm, P.-C. (1988): Umweltrecht: Einführung in ein neues Rechtsgebiet. Erich Schmidt, Berlin.

The International Board for Soil Research and Management (IBSRAM) (1997): Towards Sustainable Land Management in the 21st Century. The IBSRAM Vision, Bangkok.

The World Commission on Environment and Development (1987): Our Common Future. Oxford University Press.

Thompson, D.B. (1998): The Institutional-Transaction-Cost Framework for Public Policy Analysis. Internet, http://ssrn.com/abstract=80308, Last access: 20-8-2007.

Tiemann, S., V. Beckmann, K. Reuter and K. Hagedorn (2005): Ist der Ökologische Landbau ein transaktionskosteneffizientes Instrument zur Erreichung von Umweltqualitätszielen? [Is organic farming a transaction cost efficient instrument to achieve environmental quality targets?]. Internet, http://orgprints.org/3641/01/3641.pdf, Last access: 20-8-2007.

Tisdell, C.A. (1991): Economics of environmental conservation - economics for environmental and ecological management. Elsevier, Amsterdam; London; New York.

Toman, M.A. (1994): Economics and "sustainability": balancing trade-offs and imperatives. Land Economics 70, No. 4, pp. 399-413.

Turner, R.K. and T. Jones (1991): Wetlands, market and intervention failures. Earthscan, London.

Turner, R.K., J. Paavola, P. Cooper, S. Farber, V. Jessamy and S. Georgiou (2003): Valuing nature: lessons learned and future research directions. Ecological Economics 46, No. 3, pp. 493-510.

Umstätter, J. (1999): Calibrating regional production models using positive mathematical programming: an agro environmental policy analysis in Southwest Germany. Shaker, Aachen.

UN Department of Economic and Social Affairs (2006): Agenda 21. Internet, http://www.un.org/esa/sustdev/documents/agenda21/english/agenda21toc.htm, Last access: 20-8-2007.

Urfei, G. (1999): Agrarumweltpolitik nach den Prinzipien der Ökonomischen Theorie des Föderalismus - Ein Regionalisierungsansatz zur territorialen Abgrenzung effizienter Politikallokationsräume. Duncker und Humblot, Berlin.

Urfei, G. and R. Budde (2002): Die Geographie des Umweltföderalismus: ein empirischer Ansatz zur Bestimmung effizienter Regelungsebenen der Umwelt- und Naturschutzpolitik in Deutschland. Rheinisch-Westfälisches Institut für Wirtschaftsforschung, Essen.

Uthes, S. (2005): NCO-Produktion auf dem Naturschutzhof Brodowin - MODAM Anwendung am Beispiel der Zielart Feldhase (Lepus europaeus) - Master thesis. Humboldt Universität, Berlin.

Van den Born, G.J., B.J. De Haan, D. Pearce and A. Howarth (2000): Technical Report on Soil Degradation, RIVM report 481505018. Internet, http://www.environmental-expert.com/articles/article881/soil.pdf, Last access: 20-8-2007.

Vereijken, P.H. (2001): Country Report Netherlands. Multifunctionality: Applying the OECD Analytical Framework Guiding Policy Design, Organisation for Economic Co-operation and Development (OECD), Paris.

Walker, D.J. (1982): A damage function to evaluate erosion control economics. American Journal of Agricultural Economics 64, No. 4, pp. 690-698.

Walsh, R., J. Loomis and R. Gillman (1984): Valuing option, existence, and bequest demands for wilderness. Land Economics 60, No. 1, pp. 14-29.

Weersink, A. (2002): Policy options to account for the environmental costs and benefits of agriculture. Canadian Journal of Plant Pathology 24, No. 3, pp. 265-273.

Weersink, A., S. Jeffrey and D. Pannell (2002): Farm-Level Modeling for Bigger Issues. Review of Agricultural Economics 24, No. 1, pp. 123-140.

Weersink, A., J. Livernois, J.F. Shogren and J.S. Shortle (1998): Economic Instruments and Environmental Policy in Agriculture. Canadian Public Policy vol. 24 (3), pp. 303-327.

Werner, A. (2006): Schlaginterne Segregation - ein Konzept zur Integration von Naturschutzzielen in gering strukturierten Agrarlandschaften durch kleinflächige Stilllegungen. Leibniz-Zentrum für Agrarlandschaftsforschung, Internet, http://www.zalf.de/home_zalf/download/dir/jb_98_99/k_342ls.pdf, Last access: 20-8-2007.

Westra, J.V., K.W. Easter and K.D. Olson (2002): Targeting nonpoint source pollution control: Phosphorus in the Minnesota River basin. Journal of the American Water Resources Association 38, No. 2, pp. 493-505.

White, G. (1993): Towards a Political Analysis of Markets. IDS Bulletin 24 (3).

Williams, J.R., K.G. Renard and P.T. Dyke (1983): Epic - A New Method for Assessing Erosions Effect On Soil Productivity. Journal of Soil and Water Conservation 38, No. 5, pp. 381-383.

Williamson, O.E. (1985): The Economic Institutions of Capitalism. Free Press [u.a.], New York.

Wischmeier, W.H. and D.D. Smith (1978): Predicting rainfall erosion losses. A guide to conservation planning. Washington.

Wissenschaftlicher Beirat Bodenschutz beim BMU (2000): Wege zum vorsorgenden Bodenschutz. Fachliche Grundlagen und konzeptionelle Schritte für eine erweiterte Boden-Vorsorge. Schmidt, Berlin.

Wissenschaftlicher Beirat der Bundesregierung Globale Umweltveränderungen (WBGU) (1994): Welt im Wandel - Die Gefährdung der Böden. Economica Verlag, Bonn.

Wossink, G.A.A., A.G.J.M. Lansink and P.C. Struik (2001): Non-separability and heterogeneity in integrated agronomic- economic analysis of nonpoint-source pollution. Ecological Economics 38, No. 3, pp. 345-357.

Yang, W., M. Khanna, R. Farnsworth and H. Önal (2003): Integrating economic, environmental and GIS modeling to target cost effective land retirement in multiple watersheds. Ecological Economics 46, No. 2, pp. 249-267.

Yang, W., C. Sheng and P. Voroney (2005a): Spatial Targeting of Conservation Tillage to Improve Water Quality and Carbon Retention Benefits. Canadian Journal of Agricultural Economics/Revue canadienne d'agroeconomie 53, No. 4, pp. 477-500.

Yang, W.H. and A. Weersink (2004): Cost-effective targeting of riparian buffers. Canadian Journal of Agricultural Economics-Revue Canadienne D Agroeconomie 52, No. 1, pp. 17-34.

Yang, W., M. Khanna, R. Farnsworth and H. Onal (2005b): Is Geographical Targeting Cost-Effective? The Case of the Conservation Reserve Enhancement Program in Illinois. Review of Agricultural Economics 27, No. 1, pp. 70-88.

ZALF Müncheberg (2006): Zentrale Einrichtungen - Forschungsstation Dedelow - Aktuelles. Leibniz-Zentrum für Agrarlandschaftsforschung, Internet, http://www.zalf.de/home_zalf/institute/zentral/fs/fs/fsd/index.html, Last access: 28-8-2007.

Zander, P. (2001): Interdisciplinary modeling of agricultural land use: MODAM - Multiobjective Decision Support Tool for Agroecosystem Management. In: K. Helming (Ed.): Multidisciplinary Approaches to Soil Conservation Strategies: Proceedings; International Symposium ESSC, DBG, ZALF, Selbstverlag, Müncheberg, pp. 155-160.

Zander, P. (2003): Agricultural land use and conservation options - a modelling approach. Wageningen, Universität, Diss. 2003.

Zander, P. and H. Kächele (1999): Modelling multiple objectives of land use for sustainable development. Agricultural Systems 59, No. 3, pp. 311-325.

Die VDM Verlagsservicegesellschaft sucht für wissenschaftliche Verlage abgeschlossene und herausragende

Dissertationen, Habilitationen, Diplomarbeiten, Master Theses, Magisterarbeiten usw.

für die kostenlose Publikation als Fachbuch.

Sie verfügen über eine Arbeit, die hohen inhaltlichen und formalen Ansprüchen genügt, und haben Interesse an einer honorarvergüteten Publikation?

Dann senden Sie bitte erste Informationen über sich und Ihre Arbeit per Email an *info@vdm-vsg.de*.

Sie erhalten kurzfristig unser Feedback!

VDM Verlagsservicegesellschaft mbH
Dudweiler Landstr. 99   Telefon +49 681 3720 174
D - 66123 Saarbrücken   Fax   +49 681 3720 1749
www.vdm-vsg.de

Die VDM Verlagsservicegesellschaft mbH vertritt

Printed by Books on Demand GmbH, Norderstedt / Germany